Current Topics in Microbiology and Immunology

170

Protein Traffic in Eukaryotic Cells

Selected Reviews

Edited by R. W. Compans

With 14 Figures

Springer-Verlag
Berlin Heidelberg NewYork
London Paris Tokyo
Hong Kong Barcelona
Budapest

Professor Dr. Richard W. Compans
Department of Microbiology
University of Alabama at Birmingham
UAB Station
Birmingham, AL 35294, USA

Cover illustration by courtesy of Dr. Eric Hunter, Department of Microbiology, University of Alabama at Birmingham, USA (see p. 108, Fig. 1).

ISBN-13:978-3-642-76391-5 e-ISBN-13:978-3-642-76389-2
DOI: 10.1007/978-3-642-76389-2

Softcover reprint of the hardcover 1st edition 1991

Library of Congress Catalog Card Number 15-12910

Typesetting: Thomson Press (India) Ltd, New Delhi;

23/3020-543210 – Printed on acid-free paper.

Preface

The mechanism of intracellular transport and targeting of proteins to their final destinations is an area of great current research interest. Many newly synthesized proteins are initially directed to the rough endoplasmic reticulum, but subsequently traverse distinct transport pathways and are ultimately delivered to a variety of organelles or cell surface destinations. In this volume, the current status of our understanding of a number of these distinct transport pathways is discussed.

Mitochondria acquire their membrane and cisternal proteins by pathways which are independent of the central vacuolar system. The mechanisms of targeting of proteins to this organelle are considered in the chapter by A.L. HORWICH, M. CHENG, A. WEST, and R.A. POLLOCK.

One of the best characterized pathways for protein sorting involves the transport of protein to lysosomes. This important area of cell biology is presented by S.R. PFEFFER.

While many proteins found in the endoplasmic reticulum are transitory proteins en route to different final destinations, others remain as integral components of the endoplasmic reticulum cisternae, or the compartments of the Golgi complex. R.F. PETTERSON reviews the sorting process for such proteins, with particular emphasis on studies of viruses which are assembled at these intracellular membrane compartments.

Many proteins are transported through the central vacuolar system to the plasma membrane. D. EINFIELD and E. HUNTER discuss the structural requirements and pathways involved in this transport process.

In the case of polarized epithelial cells, the presence of tight junctions between adjacent cells serves to divide the plasma membrane into two distinct domains, the apical and basolateral domains. Distinct populations of membrane proteins, as well as secretory proteins, are transported to each of these membrane domains. The mechanisms involved in such polarized transport of proteins are discussed by R.W. COMPANS and R.V. SRINIVAS.

It is realized that only a portion of the extensive information on transport of membrane and vacuolar proteins has been presented here. It is hoped, however, that publication of this volume will provide a better understanding of the current status of research, and point out the unsolved problems to be addressed in the future.

RICHARD W. COMPANS

Contents

List of Contributors

(Their addresses can be found at the beginning of their respective chapters.)

Mitochondrial Protein Import

A. L. HORWICH, M. CHENG, A. WEST, and R. A. POLLOCK

1 Introduction

Most of the several hundred proteins that comprise mitochondria are encoded in the nuclear genome and synthesized on cytosolic polyribosomes. These proteins must be transferred from their cytosolic site of synthesis to the mitochondria, where they carry out their biological functions. A host of questions arise concerning such a traffic system. How are the newly synthesized proteins correctly targeted to the organelles? How are they translocated through the mitochondrial membranes? How are they sorted to their particular submitochondrial destinations? How are they assembled into active forms?

These questions have been a subject of intense interest and investigation during the past 10 years. Here, we review present understanding of the mitochondrial protein import pathways, devoting particular attention to mutational approaches that have advanced the comprehension of this system. Our

Department of Human Genetics, Yale University, School of Medicine, 333 Cedar Street, P.O. Box 333, New Haven, CT 06510-8005, USA

Current Topics in Microbiology and Immunology, Vol. 170

analysis of protein import is arbitrarily divided into a discussion of targeting signals within the proteins themselves and a discussion of machinery present in the organelles that must recognize these signals and act upon the proteins. At present, the molecular interactions between signals and components of the machinery remain undefined, but with rapid progress in characterization of components this will be a subject for analysis in years to come.

2 General Scheme

Mitochondrial protein import can be considered in steps that include cytosolic translation of the proteins, specific recognition by the organelles, translocation through the mitochondrial membranes, proteolytic processing of signal peptides, and assembly of polypeptides into biologically active enzymes (Fig. 1). The import pathway appears to be highly conserved through evolution. Not only have the same steps of import been observed in a multiplicity of species, but also proteins of one species have been shown to be recognized, imported, and faithfully matured by the mitochondria of another (SCHLEYER et al. 1982; TEINTZE et al. 1982; SCHMIDT et al. 1983; TAKIGUCHI et al. 1983; SCARPULLA and NYE 1986;

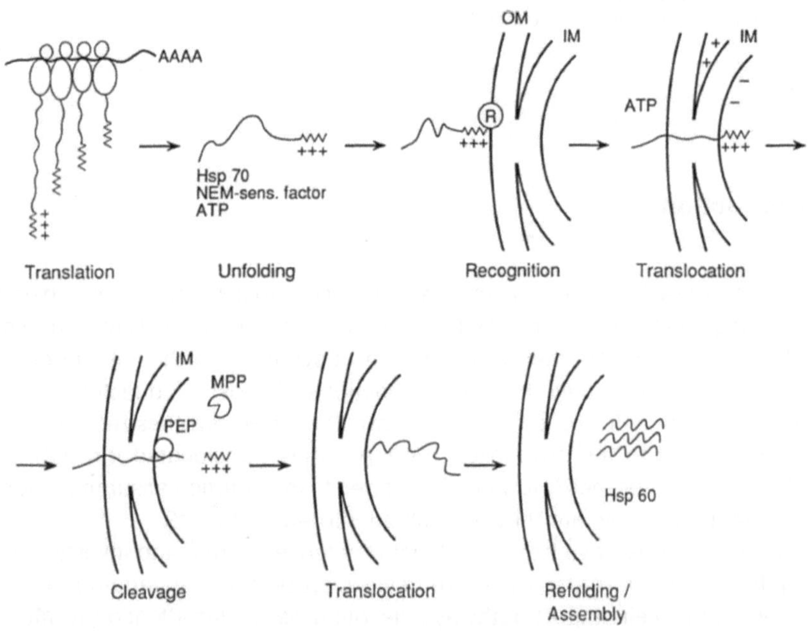

Fig. 1. Pathway of mitochondrial protein import, illustrating steps of entry of a precursor protein from the cytosol through a contact site inot the mitochondrial matrix space. *OM*, outer Membrane; *IM*, inner membrane; *MPP*, matrix processing peptidase; *PEP*, processing enhancing protein

HURT et al. 1986; CHENG et al. 1987). It appears that even if one species does not have a particular protein in its mitochondria, when programmed for expression of that protein using the coding sequence from a different species that does have the protein, its mitochondria are able to recognize and import that protein (SCHMIDT et al. 1983; CHENG et al. 1987). For example, ornithine transcarbamylase (OTC) is a cytosolic homotrimeric enzyme in *Saccharomyces cerevisiae* (URRESTARAZU et al. 1977) and a mitochondrial matrix homotrimer in man (KALOUSEK et al. 1978). Consistent with its cytosolic localization, the yeast subunit lacks a mitochondrial targeting signal (HUYGEN et al. 1987). However, when the human OTC subunit precursor was expressed in *S. cerevisiae* it was taken up by mitochondria and matured into an active matrix enzyme (CHENG et al. 1987).

Mitochondrial proteins appear to be synthesized both on free cytosolic polyribosomes and on polysomes bound to the outer surface of the organelles (KELLEMS et al. 1974, 1975; ADES and BUTOW 1980; SUISSA et al. 1984). The precise mechanisms regulating distribution of mRNAs encoding these proteins between the two populations of polysomes are unknown. While cotranslational import from organelle-bound polysomes may be operative under some physiological conditions, mitochondrial protein import can occur entirely posttranslationally. For example, a protein can first be synthesized in vitro in rabbit reticulocyte lysate or in a yeast cytosolic translation system, and then either a protein synthesis inhibitor can be added or the ribosomes removed by sedimentation. When the lysate containing the newly synthesized protein is added to isolated mitochondria, the protein is faithfully imported and matured (HARMEY et al. 1977; KORB and NEUPERT 1978; MACCECCHINI et al. 1979a, b; ZIMMERMANN and NEUPERT 1980; CONBOY and ROSENBERG 1981).

In general, proteins destined for the outer mitochondrial membrane are targeted directly to the membrane, and do not undergo proteolytic processing during integration into the membrane (FREITAG et al. 1982; HASE et al. 1983; MIHARA and SATO 1985; KLEENE et al. 1987). In contrast, proteins destined for the other compartments are translated as larger precursors containing N-terminal cleavable signal sequences called leader peptides. These amino acid sequences target the proteins to the organelles (HURT et al. 1984; HORWICH et al. 1985a; EMR et al. 1986; KENG et al. 1986), and direct them through the membranes at so-called contact sites, where outer and inner membranes come in contact (SCHLEYER and NEUPERT 1985). The leader peptides are probably recognized by receptor elements in the outer mitochondrial membrane (RIEZMAN et al. 1983; ZWIZINSKI et al. 1983, 1984) and then inserted into the outer membrane (PFALLER et al. 1988), and both these steps and the subsequent step of translocation through the membranes requires the precursor protein to be unfolded, in a "loose" conformation (EILERS and SCHATZ 1986). Both protein factors in the cytosol and ATP are required to produce such a conformation (PFANNER and NEUPERT 1986; VERNER and SCHATZ 1987; CHEN and DOUGLAS 1987a–c; PFANNER et al. 1987d; DESHAIES et al. 1988; H. MURAKAMI et al. 1988). Translocation through the inner membrane depends on an electrochemical potential gradient across this membrane normally present in energized mitochondria (GASSER et al. 1982a; KOLANSKY et al. 1982; SCHLEYER et al. 1982; PFANNER and NEUPERT 1985; HARTL et al. 1987b). During or after passage of the precursor through the contact site, the leader peptide is proteolytically cleaved (HARMEY and NEUPERT 1979; MACCECCHINTI et al. 1979a, b; MICHEL et al. 1979). Mature-size subunits entering the matrix space are then subjected to a protein-assisted folding process which assembles matrix proteins into their active forms.

It appears likely that most proteins destined for the intermembrane space utilize the matrix pathway. Following entry into the matrix and proper folding by the matrix-localized apparatus (CHENG et al. 1989), they are redirected back through the inner membrane. This second step represents a process known as conservative sorting, in which an evolutionarily ancient pathway, bacterial secretion, has been retained as a step in modern biogenesis (HARTL et al. 1986). Thus the evolutionarily new step is that of matrix targeting. We detail below what is known about both the old and the new.

3 Localization Signals

3.1 Outer Membrane Proteins

While proteins of the outer membrane lack cleavable signal sequences, these proteins are nevertheless synthesized as precursors with conformations apparently different from those of their biologically active forms. For example, porin, a 30-kDa integral outer membrane protein which serves as a passageway for ions and small molecules, is imported from a water-soluble form into the mitochondrial outer membrane (PFALLER et al. 1985). Newly synthesized water-soluble porin molecules bind to a trypsin-sensitive outer membrane "receptor" protein, followed by integration into the lipid bilayer (PFALLER and NEUPERT 1987). In this location they are completely insensitive to trypsin digestion and can be solubilized only by anionic detergents. The two states of porin are almost certainly associated with different conformations of the protein, exposing different regions to the external environment. The specific sequences involved in determining conformational changes are unknown, although it has been speculated that a hydrophobic domain between residues 23 and 39 might play a role. It is equally striking that the N-terminal half of this protein is hydrophilic in overall character, whereas the C-terminal half is extremely hydrophobic (MIHARA and SATO 1985).

The signal directing outer membrane localization of the 70-kDa protein of the yeast outer mitochondrial membrane has been experimentally defined (HASE et al. 1984, 1986). This protein, whose normal function is unknown, has been shown to be anchored to the membrane via its N-terminal 10-kDa portion, while its C-terminus resides in the cytosol. Experiments testing various deletion constructs revealed that the extreme N-terminal 12–15 residues contain information that directs mitochondrial matrix localization (HASE et al. 1984). This domain includes two lysine residues and one arginine and is devoid of acidic residues, thus resembling in amino acid composition a mitochondrial matrix targeting sequence. Additional constructs that included not only this N-terminal portion but also the adjoining residues, 13 to 38, directed correct localization of the derivative proteins into the outer membrane. These additional residues are hydrophobic in character, resembling a typical membrane-anchoring sequence. Further support for the function of two distinct domains in localization of the 70-kDa protein was obtained by experiments involving fusion of N-terminal sequences with *Escherichia coli* β-galactosidase: inclusion of only the first 21 residues of the 70-kDa protein directed the bacterial enzyme to the matrix space, while inclusion of the first 61 residues directed the enzyme to the outer membrane (HASE et al. 1986). The collective of studies of the 70-kDa Kd protein suggested that outer membrane targeting signals might simply be a variant of those directing matrix localization, differing only in that one or more hydrophobic domains are simply positioned distal to a matrix signal in order to "stop transfer." However, subsequent structural analyses of other outer membrane proteins, including porin and monoamine oxidase, fail to support this as a general mechanism. Porin is embedded by antiparallel β-pleated sheets. The porin sequence contains three acidic residues and two basic residues in its N-terminal portion, not at all consistent with the composition of a typical matrix localizing signal (MIHARA and SATO 1985; KLEENE et al. 1987). Similarly, the N-terminal structure of cytochrome c heme lyase bears no resemblance to the topogenic domains of the 70-kDa protein. The N-terminus of monoamine oxidase contains an abundance of both acidic and basic residues, and here the longest stretch of noncharged residues is only 14 in length (HSU et al. 1988). Thus, while the 70-kDa protein employs an N-terminal "stop matrix transfer" strategy to achieve outer membrane localization, other proteins of the outer membrane pursue apparently different strategies to achieve their localization.

3.2 Matrix Proteins

In contrast with the intrinsic localization signals of outer membrane proteins, the proteins destined for the mitochondrial matrix space contain N-terminal cleavable leader peptides, comprised of 20–60 amino acids. In general, these peptides contain both necessary and sufficient information to direct matrix localization, as demonstrated by gene fusion experiments. The first such studies joined signal peptides from yeast cytochrome oxidase subunit IV (COX IV) (HURT et al. 1984) or the subunit precursor of human OTC (HORWICH et al. 1985a) to the small cytosolic protein dihydrofolate reductase (DHFR). Both signal peptides directed DHFR into the mitochondrial matrix both in intact cells and in vitro. When regions within these two leader peptides were examined for targeting function, the first 12 residues of the COX IV leader (normally 25 residues) (HURT et al. 1985) and the middle 15 residues of the OTC leader (normally 32 residues) (HORWICH et al. 1987) were found sufficient to direct matrix localization. Additional fusion experiments employing other leader peptides have been carried out during the last few years. As few as three to five residues of the Heme 1 leader were sufficient to direct matrix localization (KENG et al. 1986). However, this and other signals are normally constituted in larger segments, probably to enable greater efficiency of targeting (EMR et al. 1986). For example, the terminal portions of the OTC leader peptide were dispensable to matrix localization of the subunit, but in their absence, the extent of import and cleavage of the OTC precursor by isolated mitochondria was significantly diminished (HORWICH et al. 1986, 1987). The most extreme example to date of cooperation between sequences in providing mitochondrial matrix signal function, to the extent of signal redundancy, is the yeast F1BATPase subunit (BEDWELL et al. 1987). Here, three different short sequences, amino acids 5–12, 16–19, and 28–34 are each capable of directing localization to the matrix space. Remarkably, the last of these is located in the mature portion of the precursor. Deletion of any one of these sequences does not measurably affect import. Only when all three are deleted is import function lost. Why FIBATPase evolved such a "backup" system for targeting is unclear. While the COX IV and OTC leader peptides do not appear to contain redundant matrix targeting elements, it remains to be determined whether mitochondrial signal peptides commonly contain such redundancy.

In considering matrix targeting domains the question arises as to how these signals arose in the first place. If one accepts an endosymbiotic view of mitochondrial origin, then one must conclude that as genes moved in the ancestral eukaryotic cell from the mitochondrion to the nucleus, the products, now cytosolically synthesized, must have needed to acquire signals in order to localize to the organelles. It was perhaps mechanistically most expedient to append N-terminal signal sequences. This could arise most straightforwardly either by converting 5′ nontranslated sequences into new N-terminal coding sequences or by splicing new exons to the 5′ termini of the original coding sequences. The latter mechanism, the addition of a new translated exon, is consistent with apparently ongoing processes of exon shuffling that bring various functional domains together. Yet, was intron-exon organization present at the time of endosymbiosis? A recent analysis suggests that such organization *was* present: the intron–exon boundaries of glyceraldehyde phosphate dehydrogenase genes in the plant *Arabidopsis thaliana* were found to be precisely conserved between the chloroplast genome and the nuclear genome (SHIH et al. 1988). Consistent with an exon-shuffled origin of at least some leader peptides, several mammalian leader peptides reside in individual exons. The first exon of both rat (TAKIGUCHI et al. 1987) and human (HATA et al. 1988). OTC genes encodes the N-terminal 26 residues of the precursor, sufficient to direct mitochondrial localization. The first exon of the rat carbamyl phosphate synthetase I gene encodes the entire leader peptide plus four amino acids from the mature subunit (LAGACE et al. 1987), also sufficient to direct mitochondrial localization (NGUYEN et al. 1986). The chicken δ-aminolevulinate synthase gene has a similar structure (MAGUIRE et al. 1986). In contrast, in other genes leader coding sequences lie adjoined in the first exon with large blocks of sequence of the mature portion. Other mechanisms of origin may have been utilized in these cases.

Whatever the mechanism of origin, the N-terminal position of mitochondrial matrix targeting signals is apparently without exception. Indeed, recent examination of a matrix enzyme, ketoacyl CoA thiolase, that is unusual for its lack of a cleavable signal peptide reveals that even in this protein the matrix targeting signal lies at the N-terminus (ARAKAWA et al. 1987; AMAJA et al. 1988). While one can postulate mechanisms by which matrix targeting signals arose at the N-terminus of precursors, there may be functional reasons for evolution of the signals at this location. First, study of translocation intermediates reveals that the N-terminus of precursor proteins proceeds first through the mitochondrial membranes. There may simply not be "room" in the translocation "passageway" to accommodate translocation of a recognized loop structure, thus excluding the signal on functional grounds from the interior position of a precursor. Second, in relation to the mature moiety, it may have been much easier at the time of endosymbiosis to add a cleavable signal domain to a terminus than to place it internally. An internally added signal sequence might interfere with the original catalytic activity of the polypeptide. External placement with subsequent proteolytic removal would allow the mature protein to proceed with its biogenesis precisely as if it had been translated inside the ancestral bacterial cell. In support of the notion that external placement minimizes functional interference, when leader peptides have been artificially joined at the N terminus of proteins such as DHFR (EILERS and SCHATZ 1986), β-galactosidase (DOUGLAS et al. 1984; BRANDRISS and KRYZYWICKI 1986) or metallothionein (CHEN and DOUGLAS 1987c), these passenger sequences have retained their biological activity. The ease with which matrix targeting domains might have been appended to existing coding regions has been demonstrated by the addition of short, randomly selected pieces of *E. coli* DNA to the mature coding domain of yeast COX IV (BAKER and SCHATZ 1987). Approximately 3% of the segments tested could rescue a COX IV-deficient yeast strain. When sequences of these segments were determined, they were found to exhibit the features of matrix signals described below. While their efficiencies of matrix targeting were low compared with known leader peptides, the segments in this context had sufficient targeting activity to transfer a biological function into the matrix compartment. As a corollary observation, when an internal domain of the cytosolic protein DHFR was joined to the N-terminus of intact DHFR it could direct the fusion protein to the mitochondrial matrix with low efficiency (HURT and SCHATZ 1987). Thus, when some internal sequences are placed at the N-terminus of a protein they become able to direct matrix localization. Such "cryptic" sequences, located within many coding domains, could have provided the primordial substrates from which leader coding sequences could have arisen during endosymbiosis, by a mechanism of exon shuffling.

While many naturally occurring passenger domains exhibit biological activity even while the corresponding leader peptide is attached, this is not a universal feature. The OTC precursor, for example, is incapable of assembly into the homotrimeric active enzym while the leader peptide is attached (KALOUSEK et al. 1984; ISAYA et al. 1988). In teleologic terms, there is no reason why a passenger domain should be required to be biologically active in the precursor protein.

While the attached leader peptide can influence the function of the mature domain, the converse is also observed. For example, the mature portion of the precursor of subunit 9 of the F1/F0 ATPase of *Neurospora* enhances the binding of this precursor to mitochondria (PFANNER et al. 1987c). A negative influence was observed when β-galactosidase was fused behind the F1βTPase leader peptide: the chimeric precursor was unable to reach the matrix compartment unless at least 169 F1βATPase residues were present at the N terminus, presumably because the passenger moiety interfered with exposure of the leader peptide in shorter constructs (EMR et al. 1986). When a different passenger sequence, the invertase sequence, was joined in place of β-galactosidase, the chimeric subunit could be efficiently imported when only the leader peptide of F1β was present (EMR et al. 1986). Thus the mature passenger moiety must accommodate the step of membrane translocation. Indeed, some proteins of mitochondria are apparently either too bulky or too hydrophobic to transit the membranes from the cytosolic compartment—e.g., COX I—and these coding sequences have remained in the mitochondrial genome, enabling assembly of these proteins from inside (VON HEIJNE 1986c).

3.3 Functional Elements Within Matrix Targeting Signals

Leader peptides are striking for a characteristic amino acid composition that is rich in basic arginine and lysine residues while essentially devoid of acidic residues. Domains within leader peptides that contain sufficient information for targeting inevitably contain two or three basic residues. These residues confer a net positive charge to the signal region, and they are essential for targeting function. Mutational replacement of these residues with charge-neutral residues or replacement of charge-neutral residues with acidic residues leads to either reduction or complete loss of import function (HORWICH et al. 1985b, 1986, 1987; CHU et al. 1987a, b; BEDWELL et al. 1989). The locus of function of positively charged residues remains to be determined. Possibilities include: the cytosol, where leader peptides could potentially interact with specific cytosolic factors that recognize the positive charges; the outer membrane, where a receptor component(s) might recognize the leader peptides; and the contact site, through which translocation occurs. Interaction at the last site, like the others, could involve protein–protein interaction, but here an electrostatic interaction is also possible, in which the positively charged leader peptide travels down an electrochemical potential gradient across the inner mitochondrial membrane that is relatively negative at the matrix aspect. While the presence of net positive charge is shared by virtually all matrix targeting signals, the presence of basic residues (and absence of acidic residues) at the N terminus of a protein is not alone sufficient to direct matrix localization. When several artificially designed leader peptides rich in basic residues and devoid of acidic amino acids were joined with various passenger sequences they failed to direct mitochondrial localization (ALLISON and SCHATZ 1986; HORWICH et al. 1987).

The additional requirement for mitochondrial matrix localization appears to reside in secondary structure of the leader peptides. While leader peptides share no obvious primary amino acid sequence, in many cases their collected amino acids exhibit high α helix-forming potential (VON HEIJNE 1986b). Mutational study of the OTC leader peptide provided experimental support for a functional α helix in the critical midportion of this leader peptide. When an arginine at position 23 was substituted, the extent of import and cleavage of the OTC precursor by isolated mitochondria directly correlated with the amount of α helix potential of the substituted residue—substitution with glycine led to complete failure of import by isolated mitochondria and inability to be cleaved by matrix fractions containing the processing protease (HORWICH et al. 1986). If primary amino acid sequences of leader peptides are arrayed in helical wheel plots, the residues in many cases arrange with the basic, positively charged, residues on one aspect of the wheel and noncharged, hydrophobic, residues at the opposite aspect. This α-helical character has been termed amphiphilic (VON HEIJNE 1986b), and this property may be critical for recognition of the leader peptide by the mitochondrial machinery. Alternatively, amphiphilicity may be critical for a direct interaction of the leader peptide with the membranes themselves. In relation to the latter possibility, studies with two different synthetic leader peptides that exhibit amphiphilicity indicate strong surface active properties. Synthetic peptides from the rat OTC leader (EPAND et al. 1986) and from the yeast COX IV leader (ROISE et al. 1986) could insert spontaneously from an aqueous solution into phospholipid monolayers. In the presence of detergent micelles both peptides exhibited α-helical properties. An overproduced COX IV-DHFR fusion protein was also examined (ENDO and SCHATZ 1988). Interestingly, this protein exhibited membrane-perturbing properties only when denatured, suggesting that unfolding of a precursor protein is important to allow exposure of its leader peptide, permitting interaction between the peptide and the mitochondrial membranes.

While it seems clear that an exposed leader peptide could interact directly with the lipid bilayer, there is no evidence that this does in fact occur. Rather, evidence mitigates somewhat against such an interaction. Two recent studies examining translocational intermediates, subunits in the process of traversing the membranes, reveal that these molecules can be recovered in aqueous solution by extraction with urea or alkaline pH, suggesting that, at least while in the process of translocation, the mature domain of a precursor resides in an aqueous rather than a lipid environment (PFANNER et al. 1987a). This conclusion is further supported by an additional experiment examining an artificially modified mature domain: when the highly charged, "membrane-impermeant" organic anion, stilbene disulfonate, was added to the C terminus of a COX IV-DHFR fusion protein, it failed to interfere with translocation of the fusion protein into the matrix compartment (VESTWEBER and SCHATZ 1988a). Thus, the foregoing studies suggest that at least the mature domains of precursor proteins pass through a hydrophilic, probably protein-lined channel. Whether the leaders do so as well remains to be addressed.

While many leader peptides are predicted to form amphiphilic α helices, a number of naturally occurring and artificial leader peptides do not exhibit an obvious amphiphilic character in helical wheel analyses (ALLISON and SCHATZ 1986). For example, mutations have been generated that insert positively charged residues into the noncharged face of a putative α-helical domain in the OTC leader peptide—these mutations had no effect on import of the precursor (HORWICH et al. 1987). These observations question whether amphiphilic properties are indeed operative in leader peptides. One response that accounts for the experimental results while preserving a role for amphiphilicity is that if one accepts a measure of flexibility in positioning of amino acid side chains, then amphiphilic character could nevertheless be conferred (GAVEL et al. 1988).

Recently, additional experimental support has been supplied for a functional role of amphiphilic character in at least the F1BATPase leader peptide by the "saturation" mutagenesis study of BEDWELL et al. (1989). One portion of the F1BATPase leader peptide that contains sufficient information to direct matrix localization was examined. Only two among a large number of types of mutation consistently reduced mitochondrial import function—loss of net positive charge, and substitutions predicted to affect the putative α helix, either diminishing its predicted amphiphilic character or "breaking" it.

While α-helical structure and amphiphilicity may be the functional secondary structural elements present in leader peptides, there may be other, as yet poorly defined functional secondary structural elements. These are suggested by amino acid substitutions, produced in several different leader peptides, that are predicted neither to disrupt an α helix nor to diminish net positive charge (HORWICH et al. 1987; CHU et al. 1987a; BEDWELL et al. 1989). Here, one can only speculate that such residues confer additional structural features that may be critical to specific interactions with components of the import machinery. In order to define these interactions, three-dimensional analysis of both signals and interacting components may be required.

3.4 "Complex" Signals: Targeting of Proteins to the Inner Mitochondrial Membrane and Intermembrane Space

Two general mechanisms appear to be utilized in the localization of proteins to the inner membrane and intermembrane space: one involves halting a protein during its passage toward the matrix space, the other involves first directing a protein to the matrix compartment then redirecting it back into or across the inner membrane. The latter strategy may seem unnecessarily complex, but from the vantage point of endosymbiotic evolution, the second pathway involves the simple addition of the evolutionarily "newer" step of matrix localization from the cytosolic compartment to an "old", conserved step of prokaryotic secretion.

The evidence for use of "stop transfer" localization is inferential. NGUYEN et al. (1988) inserted a 19-residue hydrophobic segment derived from the vesicular stomatitis virus (VSV) G protein into two different positions in the OTC precursor. When the sequence was placed near the C terminus of the mature part, the protein became anchored in the inner membrane with its N-terminal domain localized within the matrix. This localization was dependent upon an electrochemical potential gradient across the inner membrane, indicating utilization of the matrix pathway, and was accompanied by leader peptide cleavage, apparently mediated by the matrix processing protease. It seems that the protein became halted during transfer toward the matrix, somehow exiting the translocation pathway to enter the inner membrane. Interestingly, when the same hydrophobic segment was placed immediately distal to the OTC leader peptide, the protein localized to the outer membrane with its noncleaved leader peptide outside the organelles and mature portion within the intermembrane space. This localization did not depend on presence of an electrochemical potential gradient, and the behavior resembled that of outer membrane proteins. Presumably the complex N-terminal signal of this construction assumes an extended loop conformation to give the protein an inverted topology. Why other signals with combined matrix targeting and hydrophobic domains fail to do the same is unclear. Perhaps the shorter hydrophobic stretches in such signals are not decoded into loop extension. In any case, the topological information encoded in the natural signals is clearly different.

More complicated than the foregoing artificial constructions are the natural signals directing mitochondrial localization of such integral inner membrane proteins as the ATP/ADP carrier protein (AAC) and the uncoupler protein (UCP) of brown fat. These proteins may utilize a stop transfer strategy. Analysis of AAC, a noncleaved protein, revealed that its first 111 residues could direct DHFR to mitochondria (SMAGULA and DOUGLAS 1988), but similarly a C-terminal segment, residues 104–313, could target to mitochondria (PFANNER et al. 1987b). Clearly, multiple targeting signals are present in AAC. Because amino acids 1–72 failed to direct DHFR to mitochondria while 1–111 succeeded, and because residues 71–97 are predicted to form an α helix while residues 98–111 are predicted to form an amphiphilic positively charged region, at least one targeting signal is suggested to resemble a matrix targeting signal. The precise mechanics of topogenesis of this protein remains to be determined.

Inner membrane localization of UCP is equally mysterious. This protein, also not cleaved during its biogenesis, is presumed to assume a topology in which its N and C terminal lie in the intermembrane space while three pairs of amphiphilic α-helical transmembrane segments each cross over to the matrix space and back. In this protein, residues 13–105, containing the first α helix pair, were alone sufficient to target a passenger sequence to the inner membrane. Residues 102–307 could also direct mitochondrial localization but membrane insertion did not occur (LIU et al. 1988). Thus, while multiple signals may also contribute to mitochondrial localization of UCP, insertion into the inner membrane may proceed in an ordered fashion. Such a sequence of topogenic events is reminiscent of the insertion of polytopic proteins, like bovine opsin, into the endoplasmic reticulum (ER) membrane (AUDIGIER et al. 1987).

A further probable situation of stop transfer at the inner membrane is exemplified by two of the subunits of cytochrome oxidase, COX IV and COX V. The mature COX IV subunit is an inner membrane protein, yet its leader peptide has the features of a matrix-targeting signal, with no identifiable hydrophobic stretch (MAARSE et al. 1984). Indeed, when the COX IV leader is fused with DHFR, it directs that protein into the matrix space (HURT et al. 1984). One must presume, therefore, that the sequences localizing the COX IV subunit to the inner membrane lie within its mature portion. No hydrophobic stop transfer domain has been detected in this portion of the protein, and it thus remains unclear how this protein is halted. A similar behavior has been observed for the yeast COX Va subunit, also a protein of the inner mitochondrial membrane (CUMSKY 1985). Interestingly, when other leader peptides, including matrix targeting signals, were joined to the mature portion of the COX Va precursor, the subunit localized to the inner membrane and was biologically active (CLASER et al. 1988).

The strategy of "conservative" sorting, involving initial translocation to the matrix space followed by exit from the matrix space back through the inner membrane into the intermembrane space, has been elegantly uncovered by HARTL et al. (1986). They first examined the Rieske iron–sulfur protein (Fe/S), a protein of the intermembrane space. The precursor of this protein contains a complex leader peptide with an N-terminal 24-residue matrix targeting signal followed by an eight-residue segment that exhibits uncharged but hydrophilic character. The two portions are cleaved in successive steps. The production of the intermediate-sized form was shown to be catalyzed by the same matrix-processing protease that cleaves other matrix targeting signals, and the intermediate form was shown to reside within mitochondria. When the mitochondria were treated with digitonin at concentrations that solubilize the outer membrane and expose the intermembrane space, the intermediate form remained insensitive to proteinase K, indicating its localization within the matrix space. If precursor and intermediate Fe/S subunit were first accumulated in the matrix space in the presence of inhibitors of the matrix protease and then released from inhibition, mature subunit was produced. Digitonin extraction and proteinase K treatment revealed localization of the mature subunit in the intermembrane space. This second step of translocation, redirection of intermediate Fe/S back through the inner membrane, is remarkably analogous to the transit of the homologous Fe/S subunit precursor of the photosynthetic bacterium *Rhodopseudomonas sphaeroides* through the bacterial inner membrane (GABELLINI et al. 1985). This similarity suggests the utilization in mitochondria of a sorting scheme that is in part conserved from the ancestral bacterial origin (HARTL et al. 1986). Thus import of proteins to the intermembrane space, and perhaps the inner membrane as well, is proposed to comprise the joining of two pathways. The first step, the targeting of proteins from cytosol to matrix compartment, represents an evolutionarily recent development arising with endosymbiosis. The second path, redirection of proteins across the inner membrane, essentially preserves the ancient pathway of bacterial secretion. This hypothesis of conserved sorting predicts that the mitochondrial machinery that recognizes intermediate-sized proteins in the matrix compartment and mediates their translocation through the inner membrane will be related to the secretory elements present in bacteria. Such elements as trigger factor (CROOKE et al. 1988; LILL et al. 1988), Sec A (CUNNINGHAM et al. 1989; LILL et al. 1989), Sec B (COLLIER et al. 1988), and Prl A (ITO et al. 1983), identified in *E. coli*, should have homologs in the mitochondria.

The utilization of a conservative sorting approach extends beyond the Fe/S protein. HARTL et al. (1987a) have more recently demonstrated the same type of two-step sorting mechanism for two cytochromes that reside in the intermembrane space, cytochromes b2 and c1. The second part of the leader region of these proteins exhibits hydrophobic features resembling those of bacterial export signals, employed here for translocation across the inner membrane. Indeed this pathway of matrix import and retranslocation through the inner membrane may be the general route to the intermembrane space. It may also be utilized in the biogenesis of inner membrane proteins, though this remains to be addressed. Studies in years to come should identify and characterize machinery involved in determining whether a protein leaves the contact site on the way to the matrix space or targets back out of the matrix compartment.

4 Machinery

4.1 Cytosolic Factors: hsp 70, NEM-Sensitive Protein, Other Proteins, ATP

An increasing array of data indicate that components in the cytosol play a critical role in the import of newly translated mitochondrial precursor proteins. Perhaps the single strongest demonstration of such a requirement is the finding that a yeast mutant deficient in cytosolic hsp 70 proteins failed to process the F1BATPase subunit precursor to its mature size (DESHAIES et al. 1988). This mutant was also shown to be defective in translocation of pre-pro-α factor into the ER. Presumably both types of precursor protein fail to be taken up from the cytosol. Consistent with a critical role for hsp 70 in the cytosol, the gene family encoding these proteins has been shown to be essential in yeast (WERNER-WASHBURNE et al. 1987). While these proteins play a critical role at normal temperature, in keeping with their name they are also induced by heat stress, apparently to protect the cell. In this regard, the class of hsp 70 proteins has been suggested to play a role in maintaining biologically active conformations of proteins. PELHAM (1986) has suggested that this may be accomplished by association of hsp 70 with hydrophobic domains exposed at the surface of proteins by denaturing stresses. Somehow binding of hsp 70 to such surfaces adjusts conformation of the protein to a biologically active form. Binding of ATP then releases hsp 70.

A more direct demonstration of the role of cytosolic hsp 70 in mitochondrial protein import is provided by the recent studies of Blobel and coworkers (H. MURAKAMI et al. 1988). They exploited the longstanding observation that when mitochondrial precursor proteins are translated in wheat germ translation mixtures they fail to be taken up from those mixtures by either isolated mitochondria or isolated salt-washed microsomes. They reasoned that the factors able to mediate these reactions in reticulocyte lysate must be either inactive or missing altogether in the wheat germ lysate. However, if a postribosomal supernatant from yeast was added, the import reaction could proceed. The authors were able to distinguish two rescuing activities in the postribosomal Supernatant, one mediated by hsp 70 proteins and the other by an NEM-sensitive factor. Whether this second factor is the same as that identified as cooperating with hsp 70 in mediating protein translocation into microsomes (CHIRICO et al. 1988) remains to be determined.

A further study employing purified radiolabeled F1BATPase subunit precursor as a substrate for import into isolated yeast mitochondria demonstrated that a 40-kDa cytosolic component from yeast or reticulocyte lysate produced considerable stimulation of import (OHTA and SCHATZ 1984). Studies in mammalian systems have also implicated a role for cytosolic proteins in import. ARGAN et al. (1983) demonstrated a role for a factor in reticulocyte lysate in the import of the rat OTC precursor into isolated mitochondria. This factor was retained in a Sephadex G25 column. Measuring import of a radiolabeled synthetic leader peptide from ornithine aminotransferase, ONO and TUBOI (1988) found that a trypsin-sensitive high-molecular-weight fraction (greater than 200 kDa purified from reticulocyte lysate could stimulate import. This factor(s) was shown to be able to bind to the synthetic peptide, and could thus play a role in recognition and/or delivery of a precursor to the mitochondrial receptor apparatus. Similarly, K. MURAKAMI et al. (1988) utilized recombinant rat OTC precursor as a substrate for import and also demonstrated a requirement for reticulocyte lysate. A small protein appears to be involved in this enhancement of OTC import (M. Mori, personal communication). In a further line of investigation, examination of a yeast mutant, cyc 2, suggests that import of apocytochrome c, which proceeds through a pathway different from other mitochondrial proteins, may also require an import factor. The cytosolic half-life of apo-iso-1-cytochrome c in the cyc 2 mutant is drastically reduced, whereas that of the minor species, apo-iso-2-cytochrome c, is unaffected (G. Schlichter, M. Dumont, and F. Sherman, unpublished work). The cyc 2 gene product could be either a cytosolic component or, perhaps, a receptor element.

Establishment of the precise role of the identified factors awaits their further structural and functional characterization, but potential cytosolic contributions include: production or maintenance of precursor proteins in a conformation that enables interaction with the mitochondrial machinery; specific recognition of the leader peptide, with specific entry of precursors into the import machinery actually commencing not at the outer membrane, but rather in the cytosol; and cytosolic conveyance of precursors through the cytosol to the outer membrane. A precedent for functions of recognition and conveyance is supplied in the secretory pathway of mammalian cells, where the signal recognition particle appears to be involved with recognition of N-terminal secretory signals of many proteins, and with their delivery to the ER membrane (WALTER and LINGAPPA 1986). Whether a similar multiprotein complex might be involved with recognition and targeting of mitochondrial precursor proteins remains to be seen. However, tantalizing support comes from the observation that when newly synthesized precursor proteins are subjected to size analysis, in all cases examined so far they exhibit a greater-than-monomer size (ZIMMERMANN and NEUPERT 1980; KALOUSEK et al. 1984; CHEN and DOUGLAS 1987b; VERNER and WEBER 1989). Whether the precursors are present in protein complexes that are physiological structures, productive of import, or whether they are nonphysiologic aggregates is unclear.

CHEN and DOUGLAS (1988) provide perhaps the best evidence that the complexes may be physiological. They examined the complexed form of F1BATPase subunit synthesized in reticulocyte lysate, using gel filtration, and observed that the newly synthesized subunit migrated in a fraction corresponding to approximately 200 kDa molecular size. They suggested that this might comprise a tetramer of F1B precursor molecules and found that when an internal sequence between residues 122 and 144 was deleted the subunit migrated at much greater size. Because both the wild-type F1B precursor and the deleted precursor were efficiently imported, it appeared that both of these aggregates were productive of import. Chen and Douglas speculated that the "tetramer" complex might normally be in equilibrium with the large aggregate. Interestingly, if the leader peptide was deleted from the internal deletion construct, the subunit was no longer importable and it migrated as a monomer. This indicates that the leader is required for formation of the large aggregate but leaves open the question of whether leader function is directed by elements in the aggregate.

Whatever the components and functions of cytosolic protein factors, recent studies make it clear that precursors must assume a "loose" or unfolded conformation in order to be imported. It is presumed that at least partial unfolding is required in order to permit translocation through the membranes. This was elegantly supported by the observation that stabilization of the folded structure of a mature DHFR domain of a COX IV leader-DHFR fusion protein, by addition of methotrexate, blocked the import of the protein by mitochondria (EILERS and SCHATZ 1986). Similarly, stabilization of an F1B leader-copperthionein chimeric precursor with copper prevented its import (CHEN and DOUGLAS 1987c). Conversely, destabilization of the structure of a precursor allowed import to proceed more efficiently: when the COX IV-DHFR precursor was denatured with urea (EILERS et al. 1988), or when a set of amino acid substitutions was introduced into its mature domain (VESTWEBER and SCHATZ 1988b), import proceeded more efficiently. How this loose conformation might be mediated in the intact cell is unclear—hsp70 in the cytosol, or perhaps a protein(s) on the mitochondrial surface, could be involved. Schatz and colleagues (EILERS et al. 1988) have shown that when a COX IV leader-DHFR fusion protein is bound to energized mitochondria in the absence of ATP it is sensitive to exogenously added trypsin, compared with trypsin insensitivity of the newly synthesized fusion protein. This suggests that at least partial unfolding must occur before the mature portion of the precursor is translocated. Consistent with this idea, the bound fusion protein could be chased into the mitochondria by addition of exogenous ATP, presumed to play a role in unfolding of the mature portion or in release of the unfolded mature portion from a binding site on, for example, a protein like hsp 70. Taking a similar experimental approach, CHEN and DOUGLAS (1987a) observed that in vitro-synthesized F1B subunit precursor was bound but not imported by isolated yeast mitochondria in the absence of external ATP. Addition of a nucleoside triphosphate (NTP) enabled subsequent import of the bound material. Apparently phosphodiester hydrolysis was required, because the precursor was not imported when a nonhydrolyzable ATP analogue was added. A very similar observation was made for translocation of the cytosolic precursor of ADP/ATP translocator (PFANNER et al. 1987d). Additional support for the role of ATP in unfolding derives from a clever experiment in which elongation-arrested COX IV-DHFR fusion proteins were found to be importable by energized mitochondria in the absence of ATP. The apparent lack of tight folding of these truncated products relieved them of the requirement for ATP for entry, arguing for a role for ATP in the process of unfolding the precursor (VERNER and SCHATZ 1987). Similar observations were made with F1B precursor molecules: the wild-type precursor required externally added ATP in order to be imported. ATP was required both to enable the leader peptide to enter the matrix space and to complete translocation of the mature portion of the subunit into the matrix compartment (CHEN and DOUGLAS 1988; EILERS et al. 1987; PFANNER and NEUPERT 1986; PFANNER et al. 1988). Chen and Douglas examined the internally deleted F1B precursor that did not form a tetramer-sized complex in reticulocyte lysate and found that it was importable in the absence of ATP. These observations concerning F1B precursor further support the idea that newly synthesized precursors must be structurally reorganized in an ATP-dependent process in order to be imported.

The requirement for NTPs clearly varies from protein to protein, apparently dependent in each case upon structure of the precursor and upon its pathway of entry. For example, when the AAC was examined, its import was, like F1B, NTP-dependent, but here the steps requiring NTPs involved binding of the protein to the outer membrane, requiring a low amount of NTPs, and a step of insertion of the protein into a protease-protected location in the outer membrane, requiring a high level of NTPs (PFANNER et al. 1987d). These steps of binding and insertion of AAC, in contrast with the F1BATPase precursor and COX IV-DHFR, did not require an electrochemical potential gradient across the inner membrane. The further transport of AAC into the mitochondria also differed from that of the two other precursor molecules: it did not require NTPs but only the electrochemical gradient. Additional variation in NTP requirement was observed with three different fusion proteins, each containing a different amount of the N-terminal portion of subunit 9 of the FOATPase joined with DHFR. The shortest portion required the least amount of NTP (PFANNER et al. 1987d).

NTP-dependent unfolding was required also for import of newly synthesized protein molecules into the mitochondrial outer membrane. In contrast, the water-soluble form of porin, produced by acid-base treatment of detergent-solubilized membrane-derived molecules, did not require NTPs, presumably because it was loosely folded (PFANNER et al. 1988).

4.2 Receptor Proteins

Receptor proteins are operationally defined as proteins in the mitochondrial membranes that are able to specifically bind either mitochondrial proteins themselves or cytosolic components bearing this set of imported proteins. While receptor molecules have not yet been isolated, the evidence for their involvement is abundant. Three different studies from the laboratory of Walter Neupert and one from the laboratory of Gottfried Schatz provide the earliest evidence for receptor elements. First, Neupert and colleagues (HENNIG et al. 1983) demonstrated that in the presence of the inhibitor deuterohemin, apocytochrome c from *Neurospora* could bind to isolated *Neurospora* mitochondria without progressing through the subsequent steps of normal maturation, translocation through the outer membrane and heme attachment. The investigators showed that binding was specific, i.e., that it could be competed for by apocytochrome c from other sources, and that it is reversible and saturable. Binding was also productive—when deuterohemin was removed, holocytochrome c was produced. Scatchard analysis permitted distinction of high- and low-affinity apocytochrome c binding sites. The number of high-affinity binding sites per mitochondrion was estimated at 600. In a second study, Neupert and colleagues (ZWIZINSKI et al. 1983) showed that the ADP/ATP carrier (AAC) protein could also bind to a receptor element. Here, binding was observed by obstructing the subsequent step of translocation via disruption of the electrochemical potential across the inner membrane. Binding here also was productive: the carrier protein could be imported after restoration of the gradient. In a third study (ZWIZINSKI et al. 1984), binding both of AAC and of the outer membrane protein porin was shown to be abolished by pretreatment of the isolated mitochondria with a small amount of either trypsin or elastase. Binding of ATPase subunits 2 and 9 was also demonstrated, but was sensitive only to trypsin and not to elastase pretreatment. Based on failure of large amounts of added apocytochrome c to interfere with import of either AAC or the ATPase subunits, and upon distinct protease sensitivity of import of AAC and ATPase subunits, at least three protein receptors could be postulated. A study by Schatz and colleagues (RIEZMAN et al. 1983) also supported the presence of receptor elements, showing that the cytochrome b2 precursor could bind to isolated mitochondria or to isolated outer membrane vesicles. Here also, in the absence of an electrochemical gradient, binding to a trypsin-sensitive site could be demonstrated, and was productive upon restoration of the gradient.

A further indirect demonstration of receptor proteins in mammalian mitochondria was provided by MORI et al. (1985), who showed that the noncleavable matrix protein ketoacyl CoA thiolase could compete for import of precursor proteins of OTC and medium chain acyl CoA dehydrogenase. Subsequently, GILLESPIE (1987) identified a 30-kDa outer mitochondrial membrane protein that could be crosslinked to a synthetic OTC leader peptide. This is the same size as a protein identified by PAIN et al. (1988) in the outer envelope of chloroplasts by an antiidiotype antiserum produced from an antiserum that recognizes a synthetic leader peptide of the ribulose bisphosphate carboxylase (RUBISCO) small subunit precursor of pea. Preincubation of isolated chloroplasts with this antiserum blocked uptake of in vitro-synthesized small subunit. In immunogold staining analysis, using the antiidiotype antiserum, this putative receptor molecule localized to sites of contact between outer and inner envelope membranes. Such contact sites in mitochondria have been demonstrated to be the normal site of import of precursor proteins.

Another approach to receptor elements, employing antisera, was taken by OHBA and SCHATZ (1987a, b). They produced antisera against mitochondrial outer membrane proteins and found that antibodies directed against proteins of 45 kDa could reduce import of precursor proteins. A similar approach recently taken by SOLLNER et al. (1989) has identified a 19kD protein of the outer membrane that appears to be a receptor for most precursors. Preincubation of mitochondria with anti-MOM19 antiserum blocked import of most precursors, excepting AAC and apocytochrome c. Interestingly, MOM19 is distributed over the outer surface of mitochondria, suggesting that it delivers bound proteins into the contact site.

NEUPERT and coworkers have further resolved the nature of the import pathway receptor components. When radiolabeled water-soluble porin was examined, its binding to mitochondria was found to be inhibited by pretreatment of mitochondria with trypsin (PFALLER and NEUPERT 1987). Porin molecules bound at 0 °C became only moderately resistant to treatment with exogenous proteases, while warming to 25 °C caused the molecules to become completely protease resistant. This was interpreted as evidence for a pathway that allowed the transfer of porin molecules from the binding site to a position in which they were completely inserted into the outer membrane. Similarly, intermediate steps in the binding of AAC to mitochondrial receptors were elegantly determined (PFANNER and NEUPERT 1987). In this case, when in vitro-synthesized AAC molecules were incubated with mitochondria at low temperature in the absence of ATP, the AAC molecules were associated with mitochondria but were sensitive to exogenously added proteases (the so-called stage 2). When incubation was carried out in the presence of ATP but in the absence of an electrochemical potential gradient, the AAC molecules became positioned in the outer membrane in a protease-protected form (stage 3).

Recently, it was demonstrated that water-soluble porin could compete for import of precursors destined for three other mitochondrial compartments. The porin competed with the Fe/S protein, which normally enters the matrix space and exits back through the inner membrane toreside in the intermembrane space; it competed with AAC and ATPase subunit 9, both of which apparently enter into the inner membrane; and it competed with the F1BATPase subunit, which enters the matrix (PFALLER et al. 1988). Competition was observed only when incubation was first carried out for 40 min. at 0 °C, followed by a shift to 25 °C; it was not observed when incubation was first carried out at 25 °C. Thus competition required that porin initially occupy its outer membrane binding site. In the case of F1B, the competition could be shown to occur at a step distal to interaction of F1B and porin receptor sites, via two related observations. First, binding and subsequent import of porin was found to be sensitive to elastase pretreatment of mitochondria, whereas import of F1B was not (ZWIZINSKI et al. 1984). Second, F1B import was no longer competed by porin when mitochondria were pretreated with elastase (PFALLER et al. 1988). This indicates that competition for import lies at a site distal to binding.

A similar conclusion could be drawn with competition by porin for the import of AAC. Here, porin was unable to compete for the formation of the stage 2 intermediate of AAC, the ATP- independent binding step, but porin could compete for the generation of the stage 3 membrane-inserted, protease-protected intermediate. Collectively, these data indicate that porin is able to compete with different precursors for the interaction with a common component present in the outer membrane, called the general insertion protein or GIP (PFALLER et al. 1988).

Interaction with GIP occurs at a step in the import pathway distal to the initial recognition and binding of precursor proteins to their respective receptors (Fig. 2). The requirement for ATP to enable AAC to reach the site of GIP may suggest that precursor proteins must assume an unfolded state in order to reach the point of interacting with GIP. GIP is proposed to occupy the position of a clearing house in the import pathway, from which proteins can be inserted into

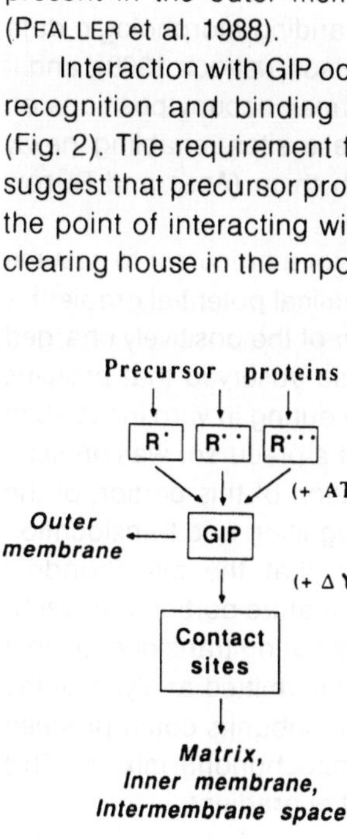

Fig. 2. A working model for the steps of recognition of precursors by receptors and insertion into the outer membrane, taken from PFALLER et al. (1988). Proteins are first recognized specifically by one or more receptor proteins, designated *R**–*R* ***, then are inserted into the outer membrane protein through a shared component, an outer membrane called the general insertion protein (*GIP*). See text for details

the inner membrane, translocated across it, or inserted into the outer membrane as in the case of porin. A candidate molecule for GIP is the 42-kDa species originally identified by OHBA and SCHATZ (1987a). They showed that antiserum raised against this protein could strongly inhibit import into trypsin-treated mitochondria (OHBA and SCHATZ 1987b). More recently this protein has been directly crosslinked with a precursor protein (VESTWEBER et al. 1989).

4.3 Translocation of Proteins Through Contact Sites into the Matrix Space

The study of translocational intermediates by SCHLEYER and NEUPERT (1985) established that proteins destined for the matrix compartment pass through sites where the outer and inner mitochondrial membranes come in contact. The intermediates were first observed by carrying out in vitro import reactions at reduced temperatures extending down to 0 °C. Under these conditions mature-sized proteins could be observed that remained susceptible to externally added proteinase K. This indicated that the leader peptide portion of the precursor had reached the matrix space and been cleaved by the matrix processing protease while the remaining mature subunit portion remained externally exposed. Additional evidence for this topology derives from the observation that specific antisera could recognize the exposed portion of such molecules (SOLLNER et al. 1988). SCHLEYER and NEUPERT (1985) reasoned that this membrane-spanning topology could only be accomplished if the mitochondrial outer and inner membranes were in contact with each other. Longstanding morphologic data indicated the presence of such physical structures (HACKENBROCK 1968), and it had even been previously suggested that ribosomes may occupy positions on the mitochondrial surface that correspond with these sites, enabling newly synthesized proteins to be imported directly through them (ADES and BUTOW 1980; KELLEMS et al. 1974, 1975).

SCHLEYER and NEUPERT (1985) found that in order for a precursor to assume a membrane-spanning topology, an intact electrochemical potential gradient is required, consistent with the notion that translocation of the positively charged leader peptide is mediated by this gradient. They also observed that proteins could be halted in this membrane-spanning topology during in vitro incubation when antibody directed against the mature portion of a precursor was present, presumably obstructing the unfolding and translocation of this portion of the precursor. Interestingly, this did not interfere with recognition and translocation of the leader portion, further supporting the idea that the mitochondrial machinery can interact with the leader even when the mature portion is covered with bulky antibody molecules. The halting of subunits during translocation in a position where only the leader has been translocated permitted analysis of the requirements for translocation of the mature portion. Subunits could proceed into the matrix space following simple warming of the mitochondrial mixture. This step did not require an intact electrochemical potential gradient.

Neupert and coworkers (SCHWAIGER et al. 1987) have further investigated the nature of the contact sites and found that they are largely resistant to digitonin treatment, which removes the outer membrane; that is, when translocation intermediates were first formed by preincubation of F1B precursor in the presence of antiserum, they cofractionated with the mitoplasts following digitonin treatment. Consistent with retention of contact sites after digitonin treatment, isolated mitoplasts were found to be able both to form contact site intermediates of F1B and to import the protein. However, if the mitochondria from which mitoplasts were prepared were first treated with trypsin, the mitoplasts became unable to form intermediates or to import F1B. This indicates that outer membrane proteins are involved in the import reaction observed with mitoplasts and that the inner membrane alone cannot mediate import. In additional experiments, contact sites were detected even in the absence of a membrane potential and in the absence of precursor proteins. Also, following extraction of mitoplasts with digitonin and high salt, they became import incompetent, but addition of a factor extracted during this procedure enabled restoration of import. This factor was apparently not simply involved with maintenance of electrochemical potential, because in its absence import was blocked even when the electrochemical gradient was supported by a variety of added substrates. A further morphologic observation confirming the biochemical identification of contact sites was that when protein A gold particles were added externally to mitochondria bearing translocation intermediates formed by addition of antiserum to the mature portion of a precursor, the gold particles localized to contact sites.

The nature of the organization and precise function of contact sites awaits isolation of the component proteins. Two major approaches are currently being taken by several laboratories. One involves the purification of translocational intermediates by physical means, isolating along with such intermediates the proteins of the contact site. VESTWEBER and SCHATZ (1988c) have "jammed" into contact sites a tripartite fusion protein containing COX IV leader and DHFR followed by bovine pancreatic trypsin inhibitor (BPTI), a disulfide-bridged protein of 6 kDa. Mitochondria "jammed" with this protein could not import added precursor proteins, and they retained their electrochemical gradient, suggesting that import sites had been saturated. When a photactivatable crosslinker was placed between the DHFR and BTTI moieties, and the same conditions for jamming employed, a flash of light resulted in the crosslinking of a 42kD protein (VESTWEBER et al. 1989). This protein proves to be the same outer membrane protein identified by an antiserum previously developed (OHBA and SCHATZ 1987) The serum can now be used to isolate the corresponding gene and to determine whether the function is essential. As mentioned earlier, it seems possible that this protein could be the general insertion protein (GIP).

Neupert and coworkers have used a fusion protein consisting of the 167 N-terminal residues of cytochrome b2 precursor fused to DHFR. This construct is able to reversibly block translocation contact sites upon addition of methotrexate. Approximately 3000 molecules spanning the two membranes were accumulated per single mitochondrion at saturation, which is equivalent to occupancy of about 1% of the total outer membrane surface (RASSOW et al. 1990). Presumably these physical complexes between translocation intermediates and translocation machinery can now be isolated and characterized. A second approach to the contact site involves genetic strategies. Using such techniques, A. West and A. Horwich (unpublished results) have identified a gene in yeast encoding a 70-kDa mitochondrial protein that is hydrophobic along nearly its entire length. It may be a component of the contact site.

4.4 Proteolytic Cleavage

As revealed by the studies of SCHLEYER and NEUPERT (1985), once the leader portion of a precursor protein enters the mitochondrial matrix space, even before the mature portion enters, it can be recognized and proteolytically cleaved by the matrix processing protease. Whether cleavage normally takes place before the precursor is completely translocated into the matrix space or usually occurs after completion of translocation is not plain, but a variety of experiments involving both intact mitochondria and matrix extracts indicate that cleavage can clearly proceed in either situation (SCHMIDT et al. 1984). However, an observation supporting the idea that contranslocational cleavage can be physiological is the finding that one component of the *Neurospora* processing enzyme resides at the inner aspect of the inner mitochondrial membrane, most likely at the inside of the contact sites (HAWLITSCHEK et al. 1988). While cleavage may occur during translocation, the translocation process itself does not require cleavage in order to be completed: inactivation of the processing peptidase, either by addition of chelators or by mutational alteration, does not interfere with ability of precursors to enter the matrix space (ZWIZINSKI and NEUPERT 1983; POLLOCK et al. 1988; JENSEN and YAFFE 1988).

The processing event carried out by the processing protease can be considered both from the standpoint of the substrate, the precursor proteins, and from the vantage point of the processing protease itself. As the primary structures of leader peptides from a considerable number of precursors have been analyzed (HARTL et al. 1989; HENDRICK et al. 1989; VON HEIJNE et al. 1989), it has become clear that the cleavage specificity of this enzyme differs substantially from those both of proteases that cleave signal sequences from secretory proteins (VON HEIJNE 1986a; BOHNI et al. 1988) and of proteases that cleave propeptides from hormones (STEINER et al. 1980). The former usually cleave after a short side chain residue while the latter cleave after a basic residue. In the case of the matrix protease, cleavage was found after a variety of residues. Mutational alteration of the terminal glutamine (32) residue of the OTC leader to phenylalanine, lysine, asparagine, or glycine had no apparent effect on proteolytic processing (HORWICH et al. 1987). Thus, residues at the site of cleavage apparently do not dictate the specificity of processing by the matrix processing protease. Rather, residues remote from the cleavage site play a role.

Perhaps most striking was the observation of HURT et al. (1987) that deletion of residues 2–5 of the 23-residue COX IV leader resulted in failure of imported COX IV-DHFR precursor to be proteolytically processed. In a similar analysis, NGUYEN et al. (1987) found that while deletion of residues 22–30 from the 32-residue rat OTC leader had no effect on import, the deleted precursor could not be processed. In additional amino acid substitution studies, HORWICH et al. (1986) found that when arginine 23 in the OTC leader peptide was changed to glycine, the substituted precursor could neither be imported by isolated mitochondria nor cleaved by matrix fractions containing the processing enzyme. The additional single substitutions in the human OTC leader peptide, asparagine 24 to arginine and Arginine 26 to leucine, led to production of intermediate-sized species (HORWICH et al. 1987). Thus residues *within* leader peptides play a critical role in directing proteolytic processing, and they most likely contribute to a recognized secondary structure. This would predict that subtle changes in amino acid side chains within the leader could interfere with cleavage. This has in fact been observed. Processing of OTC produced an intermediate when arginine 26 was changed to leucine, but there was no effect when the residue was changed to tryptophan (HORWICH et al. 1987). The secondary structure required for recognition by the matrix processing peptidase may be distinct from the structure involved with recognition and/or translocation, as suggested by a study carried out by Rosenberg and coworkers. They found that a set of substitutions involving simultaneous alteration of leucine residues 5, 6, and 8 of the rat OTC leader peptide to proline or alanine led to failure of import of the substituted precursors, but the precursors could still be cleaved to mature form by matrix fractions (KRAUS et al. 1988). A converse observation was made by PILGRIM and YOUNG (1987) who examined a mutant yeast ADH III precursor protein containing a triple substitution in the leader peptide—this precursor was efficiently translocated to the matrix space in intact yeast, but could not be cleaved.

While recognition of a secondary structure, perhaps a protein of an amphiphilic α helix, appears to be involved in directing cleavage, a particular amino acid at a specific location could still be required, perhaps as part of such a structure. Neupert and coworkers (HARTL et al. 1989) have recently addressed this question, examining 39 known sites of leader cleavage. They found that in 21 cases an arginine was present at the −2 position. Because lysine was present in only one case, it seems likely that it is the side chain structure of arginine, and not its positive charge, that is the contributing feature at this position.

Recent studies have indicated that a number of matrix-bound precursor proteins are cleaved to their mature sizes in two steps. Two matrix-localized processing proteases have been partially purified from rat liver mitochondria (KALOUSEK et al. 1988), and addition of these enzymes, known as P1 and P2, in sequence to in vitro-synthesized rat OTC precursor produced first an intermediate-sized form and subsequently the mature-sized subunit. P2 cannot cleave the OTC precursor, while P1 cleaves the precursor only to an intermediate size. Studies using these enzymes recapitulate cleavage events observed when rat

OTC precursor is incubated with rat liver mitochondria, whereupon the precursor is imported and converted first to an intermediate size and subsequently to a mature size. The intermediate is produced by cleavage after residue 24, a position +2 from arginine 23 (SZTUL et al. 1987). At least three proteins that undergo two-step cleavage, rat and human OTC and rat malate dehydrogenase (MDH), clearly do not follow the −2 rule at the second cleavage step. Rather, when the intermediate-sized species of OTC and MDH have been analyzed, they appear at this first step of cleavage to conform to the −2 rule (SZTUL et al. 1987, 1988). This suggests that these intermediates were produced by the same matrix processing enzyme that produced the mature forms of at least 21 of the other precursors examined. It also suggests that a second matrix localized processing enzyme, with a different specificity, is present. Examining further the primary amino acid configuration of the three leader peptides processed in this two-step fashion, both Rosenberg and colleagues (HENDRICK et al. 1989) and VON HEIJNE et al. (1989) have discovered three shared properties: eight residues are left for cleavage by the second enzyme, named octapeptidase; the N-terminal residue of the octapeptide is a hydrophobic residue, typically phenylalanine or leucine; and the fourth residue of the octapeptide is serine, threonine, or glycine. Eighteen leader peptides other than OTC and MDH have been found to exhibit such a potential two-step cleavage motif and await experimental demonstration of two-step processing.

Precursor proteins that follow the pathway of conservative sorting, first entering the matrix space then existing back through the inner membrane, also undergo two-step cleavage (DAUM et al. 1982; GASSER et al. 1982b; OHASHI et al. 1982; TEINTZE et al. 1982; HARTL et al. 1986, 1987a; VAN LOON et al. 1986; VAN LOON and SCHATZ 1987). Here, the first processing event, mediated by the "matrix processing" enzyme, takes place in the matrix compartment, followed by a second step of cleavage in the intermembrane space. At least two different proteases can mediate the latter step, as revealed by the studies of PRATJE and coworkers (PRATJE et al. 1983; PRATJE and GUIARD 1986). They identified a temperature-sensitive nuclear petite mutant of yeast in which both the processing of mitochondrial-coded COX II and the second cleavage step of nuclear-coded cytochrome b2 are defective at nonpermissive temperature. Interestingly, the amino acid sequences at the sites of cleavage of these two proteins are similar: the former cleaved between asparagine and aspartic acid, the latter between asparagine and glutamic acid. While the mutation in this yeast strain affected cleavage of these two proteins, it did not affect cleavage of another protein that is subjected to a second cleavage step in the intermembrane space, cytochrome c peroxidase (PRATJE and GUIARD 1986). This suggests that there are at least two different processing enzymes residing in the intermembrane space compartment.

The matrix processing protease of yeast, *Neurospora*, and mammalian cells has been studied. The enzyme exhibits unique biochemical properties: it is soluble in the absence of detergents; it exhibits neutral pH optima; it is inhibited by divalent cation chelators and stimulated by cobalt, manganese, and zinc; and it is insensitive to serine protease inhibitors like phenylmethyl sulfonylfluoride (CONBOY et al. 1982; MCADA and DOUGLAS 1982; MIURA et al. 1982; BOHNI et al. 1983; SCHMIDT et al. 1984). The gene for the yeast enzyme has recently been isolated by obtaining first a mutant strain where the enzyme is defective. This approach was initiated by YAFFE and SCHATZ (1984), who isolated two complementing temperature-sensitive lethal mutants of *S. cerevisiae* that accumulated the F1BATPse subunit precursor at nonpermissive temperature. Not only intact cells but mitochondria isolated from both of these strains were found to be likewise deficient at high temperature in cleavage of precursors incubated in vitro (YAFFE et al. 1985). The distinct nature of the two encoded products was indicated by a further elegant analysis of the *Neurospora* protease by HAWLITSCHEK et al. (1988). They purified the leader processing activity and found that the final preparation contained two proteins, one a 57-kDa protein called the matrix processing peptidase (MPP), which exhibited approximately 5% of normal processing activity by itself, and the second a 52-kDa protein called the processing enhancing protein (PEP), which when added to MPP restored full processing activity but which alone exhibited no activity. MPP was found as a soluble protein of the matrix while PEP was found loosely associated with the inner membrane. Using antibody screening, a PEP cDNA was isolated and sequence analysis predicted a protein that is hydrophilic overall but which contains two hydrophobic stretches, one near the terminus and a second in the C-terminal region. Comparison of the predicted sequence with the N-terminal amino acid sequence of the purified protein indicated a 28-residue leader peptide with a typical composition—six basic residues and no acidic residues.

While roughly equimolar amounts of PEP are sufficient for complete proteolytic activity, the protein is present both in cells and mitochondria in large excess with respect to MPP. This may reflect functions in addition to enhancing proteolysis. Studies suggest that mutations affecting the yeast homolog of PEP produce not only deficiency of proteolysis of precursor proteins but also a defect in completion of translocation of precursors through the membranes: when import is carried out in vitro at 37 °C with PEP-deficient mitochondria, the precursors remain susceptible to exogeneously added protease (YAFFE and SCHATZ 1984; F.-U. Hartl, unpublished work). Thus PEP may play a role in translocation, perhaps recognizing charge or secondary structural elements in the leader peptides and facilitating their presentation to MPP for cleavage. The PEP protein of yeast is closely related to that of *Neurospora* and has been shown to be essential for cell viability (JENSEN and YAFFE 1988; WITTE et al. 1988). Its gene rescued the mutant of Yaffe and Schatz called mass 1.

Concerning its multifunctional nature, recently in *Neurospora* PEP has been shown to be identical to the subunit I protein of cytochrome c reductase, a protein that is part of a complex that mediates electron transfer (SCHULTE et al. 1989). Thus in *Neurospora* PEP has been established to have two functions. If it is also involved with translocation in *Neurospora*. It would in fact be *trifunctional* in this organism. In yeast, however, the subunit I protein is the product of a different gene, the *CorI* gene (TZAGALOFF et al. 1986), a gene distinct from that encoding PEP, whose product is related at the level of only 25% identity. Thus PEP of *S. cerevisiae* appears not to play the part in electron transfer that its *Neurospora* counterpart does, i.e., PEP appears to be *bifunctional* in yeast.

Schatz and coworkers (YANG et al. 1988) recently purified the MPP from *S. cerevisiae*, and in contrast with the study of the *Neurospora* protease, found activity only when the two subunits, MPP (51 kDa) and PEP (48 kDa), were associated. The MPP subunit of yeast corresponds to the other complementation group that was originally isolated by YAFFE and SCHATZ (YAFFE and SCHATZ 1984). This was analyzed by HORWICH and coworkers (POLLOCK et al. 1988) who independently isolated, using a scheme of mutant isolation described below, a group of mitochondrial import function (mif) mutants, one of which, called mif 2, could not complement the original mas 2 isolate of Yaffe and coworkers, indicating that it contained the same genetic defect. Mif 2 mitochondria could import precursors at nonpermissive temperature, but could not cleave them. Detergent extracts of mif 2 mitochondria could cleave precursors at 23 °C but not at 37 °C. This strongly suggested that MPP was affected in this mutant. Similar results were obtained by JENSEN and YAFFE (1988). HORWICH and coworkers (POLLOCK et al. 1988) cloned the *mif 2* gene and used its coding sequence to generate a fusion protein to which antisera were prepared. These sera specifically recognized a predicted 52-kDa protein product that localized to mitochondrial matrix fractions. Most significantly, following incubation with antiserum and protein A sepharose, the processing activity was specifically depleted from a mitochondrial extract, demonstrating that the 52-kDa protein is MPP.

The encoded MPP product was predicted to be a hydrophilic protein, three basic residues and no acidic residues were detected in the first 16 residues, suggesting that a short leader peptide might be present. Indeed, when imported in vitro, the protein underwent a slight decrease in apparent size, suggesting that cleavage does occurs after either position 9 or 16. If MPP is indeed cleaved, then MPP must cleave itself. This most likely occurs in a bimolecular fashion as opposed to an autocatalytic mechanism. That is, newly synthesized MPP is cleaved by pre-existing MPP. This is reflective of the more general observation that mitochondria are not formed de novo but rather arise from preexisting mitochondria.

Like PEP, the MPP product was also found to be essential for the growth of yeast. Surprisingly, MPP was fond to be structurally related to PEP, the two proteins exhibiting 26% identity and 48% similarity, suggesting that they may have evolved from a common ancestor. If one accepts the notion of a common origin, then it would appear that the functions of these two components of the processing enzyme represent considerable divergence, such that PEP plays roles in both enhancement of proteolysis and translocation of precursors while MPP plays a role apparently only in proteolysis. Efforts to introduce a function of one component into the other by exchanging domains have been unsuccessful to date.

Two mammalian processing proteases have recently been partially purified from mitochondrial matrix fractions and have been found to exhibit requirements differing from those of the yeast and *Neurospora* enzymes (KALOUSEK et al. 1988). The enzyme P1 corresponds in cleavage specificity to the MPP enzyme of yeast and *Neurospora*. However, this enzyme does not require divalent metal ions. Its size is approximately 50 kDa. Whether its activity is normally enhanced by a PEP-like protein is unknown. The enzyme P2, the octapeptidase enzyme, exhibits features differing from P1, including a divalent metal ion requirement and sensitivity to N-ethyl maleimide. The octapeptidase enzyme may have a homolog in the matrix space of yeast and *Neurospora* mitochondria, where it appears that in at least three cases—COX IV, the Rieske Fe/S protein, and cyclophilin—the leader peptide may normally be cleaved in two steps, the first involving MPP, the second an enzyme conforming to the octapeptidase cleavage rules (MAARSE et al. 1984; HARTL et al. 1986; TROPSCHUG et al. 1988; HENDRICK et al. 1989).

Given the primary amino acid sequence of the matrix processing protease, can one predict how this enzyme might function? BAZAN (personal communication) has carried out computer modeling studies that suggest that the MPP subunit is an aspartyl protease. The mechanism of action of this class of enzymes would suggest that an α-helical leader peptide domain would be aligned within a longitudinal groove in the enzyme, following which attack by an aspartate residue at an active site would cleave the peptide bond joined with the mature domain. This proposed mechanism would offer explanations both for why internal leader structure can influence cleavage and for why there is no requirement for a particular residue(s) at the site of cleavage. Mutational and structural analysis of MPP should now be directed at establishing whether such a mechanism is operative.

4.5 Cytochrome c Import Pathway

Import of cytochrome c proceeds via a pathway that differs from that of other precursor proteins. Other proteins follow a pathway that (1) requires ATP; (2) involves interaction with a receptor protein(s); (3) involves interaction with the GIP; and (4) requires an intact electrochemical potential gradient for translocation through the contact site. Cytochrome c does not require any of these (ZIMMERMANN et al. 1981; NICHOLSON et al. 1987, 1988; NICHOLSON and NEUPERT 1990). The recent studies by Neupert and coworkers have delineated a unique pathway of import that involves two major steps: spontaneous insertion of apocytochrome c (apo c) into the outer membrane with accompanying binding to cytochrome c heme lyase (CCHL), and translocation through the membrane associated with heme attachment catalyzed by the lyase enzyme (NICHOLSON et al. 1987, 1988).

NEUPERT and coworkers (NICHOLSON et al. 1988) distinguished the first step of apo c import by showing that when in vitro-synthesized apo c was bound to isolated mitochondria in the absence of translocation (blocked by maintaining heme in an oxidized state), it remained susceptible to exogenously added proteinase K. When the reducing agent dithionite was added to such a mixture, producing reduction of heme and enabling heme attachment, holocytochrome c (holo c) was formed and was resistant to digestion by added proteinase K. This distinguished a second step of membrane translocation. The binding step was next shown to involve production of a complex of apo c with CCHL. First, increasing concentrations of KCl were shown to inhibit both association of apo c with isolated mitochondria and the subsequent conversion to holo c upon dithionite addition. When the binding reaction was first carried out in the absence of KCl, subsequent addition of salt had no effect upon binding as reflected by ability to chase bound apo c into holo c with dithionite. Next, mitochondrial extracts were prepared using the detergent octyglucoside. These extracts contained solubilized CCHL, capable of converting apo c to holo c. When the KCl-dithionite protocol was repeated in this system the results, measured as production of holo c, were identical, strongly indicating that binding of apo c involves a direct interaction of apo c with the heme lyase. Thus in the case of binding to intact mitochondria, it appears that while bound apo c is exposed externally, it must at least in part penetrate the outer membrane to interact with CCHL. CCHL itself was a candidate for mediating the specific binding of apo c. such an idea is supported by the observation that cyt 2-1, a mutant of *Neurospora* that is defective in CCHL activity, is also defective in binding of apo c to mitochondria (NARGANG et al. 1988). Whether CCHL alone mediates binding remains to be seen, but it seems clear that binding of apo c to mitochondria is not mediated by a classical surface receptor, because pretreatment of mitochondira with amounts of proteases that disrupt import of other precursors has no effect on apo c import. Evidence for an additional protein involved in binding comes from the observation that when intact mitochondria were treated with digitonin to open the outer membrane, followed by protease treatment, CCHL activity was destroyed, but the organelles could still bind apo c in the same manner as untreated organelles: the Scatchard plot for such binding looked identical to that for binding of apo c to intact mitochondria, indicating that binding occurs both with the same affinity and through the same apparent number of sites (NICHOLSON et al. 1988). While a second protein could be involved, it seems also possible that the lyase could function through two distinct domains: one domain in the membrane could be involved with binding, and this domain could be protease resistant; a second domain could catalyze heme addition, and could be exposed at the inner aspect of the outer membrane. Finally, it is noteworthy that c has a tendency to spontaneously insert into lipid membranes (RIETVELD et al. 1983; BERKHOUT et al. 1987), raising the question of whether a protein-lipid interaction might constitute at least the initial binding step.

Remarkably, the pathway of cytochrome c import is not totally isolated, but can intersect with the receptor/GIP-mediated pathway (STUART et al. 1990). This has been revealed by the behavior of a fusion protein joining the matrix targeting portion of the cytochrome c1 leader peptide with apo c. In the presence of an electrochemical gradient but absense of NADH, leaving heme in an oxidized state, this protein entered the receptor/GIP pathway to reach the matrix. In the presence of NADH but absence of an electrochemical gradient, the chimeric protein entered the apo c import pathway to reach the intermembrane space. Most informatively, if mitochondria were pretreated with proteinase K to destroy the receptor that recognizes pre-c1, the protein could still reach the matrix if an electrochemical gradient was present. GIP was not required in this pathway, because when large amounts of water-soluble porin (which uses a distinct receptor and GIP) were added, they did not compete for import to the matrix. Remarkably, import to the matrix could also be observed in proteinase K-treated mitochondria of a heme lyase-deficient cyt 2-1 strain of *Neurospora*. Here, both the c1 receptor and cytochrome c import apparatus are ablated. Apparently the properties of the cytochrome c moiety of the fusion protein can direct membrane insertion, enabling bypass of receptor/GIP interaction. The inserted protein can them employ the electrochemical gradient to enter the matrix space. Finally, as might be predicted, when the protein was first bound to proteinase K-treated mitochondria in both the absence of reducing agent and absence of the gradient, it could be chased into either the cytochrome c or the matrix pathway by addition of reducing agent or membrane potential respectively.

4.6 Assembly of Proteins in the Mitochondrial Matrix

Perhaps the step in mitochondrial import that has received the least attention to date is that involving the assembly of mature-sized proteins into their active forms. It remained possible that, once precursor proteins had been translocated through the membranes and cleaved to mature size, they could spontaneously find the correct conformation to exhibit biological activity. However, recent genetic evidence indicates that assembly is protein assisted. CHENG et al. (1989) have isolated a temperature-sensitive lethal mutant of yeast that is defective in the assembly process; at nonpermissive temperature mitochondrial subunits reached the matrix compartment and were cleaved to their mature sizes, but they failed to assemble into their active forms.

This assembly-deficient mutant was identified from a group of mif mutants isolated by relying on one assumption and one observation: It was *assumed* that obstruction of protein import would be lethal to the cell, based on the notion that not only the respiratory chain but also steps of other essential metabolic pathways, including those involved with lipid and amino acid metabolism, reside inside mitochondria. Failure to supply active forms of such enzymes in the correct location would result in blockade of the corresponding pathways. It had also been observed that when the precursor of human OTC was expressed in *S. cerevisiae*, it was imported to the matrix space of the yeast mitochondria, cleaved to its mature size, and assembled into the enzymatically active homotrimer (CHENG et al. 1987). This observation was particularly surprising, because while *S. cerevisiae* has its own OTC enzyme, also a homotrimer of 36-kDa subunits, this enzyme is cytosolic. Consistent with this localization, there is no leader peptide encoded at the 5′ terminus of the yeast OTC sequence. The observation of import of the human OTC precursor by mitochondria of yeast suggested two conclusions: first, that the human precursor most likely utilizes a pathway generally employed by yeast precursors to enter the matrix space; and second, that the pathway of mitochondrial protein import is conserved in evolution between yeast and man.

HORWICH and coworkers (POLLOCK et al. 1988; CHENG et al. 1989) next utilized a strain of yeast devoid of endogenous OTC, into which they had inserted the human OTC precursor coding sequence joined to a galactose operon-inducible promoter. The derived strain was mutagenized with ethylmethane sulfonate, and a library of temperature-sensitive lethal mutants was isolated (Fig. 3). Each of these mutants was subjected to a "simultaneous shift assay" in which the strain was shifted from 23 °C to 37 °C and at the same time expression of OTC precursors was induced by shifting from glycerol-containing medium to galactose-containing medium. After 2h at 37 °C cells were extracted and OTC enzyme activity was measured. Thus the intactness of the mitochondrial import pathway could be easily assessed, because OTC activity could only be generated by import of OTC precursor followed by the production and assembly of mature-sized subunits. Specifically, the OTC subunit in its precursor form is not capable of assembling into a trimer and does not exhibit enzyme activity. In addition to detecting mutations affecting the import pathway, the genetic screening would also detect mutations affecting transcription, translation, or galactose induction. While these presented a degree of "noise" to the screen, the ability to carry out hundreds of enzyme assays in a few hours made this technique very attractive. Literally thousands of temperature-sensitive lethal mutants were screened over a period of several weeks. Only a few percent failed to produce activity in the shift assay, a manageable number for the next step of the screening, which was designed to distinguish the various loci affected. This step involved immunoblot analysis carried out on extracts from the shifted cells using both anti-OTC antisera and anti-F1BATPase antisera. This uncovered essentially three classes of mutant: one that accumulated precursors, one that produced mature-sized subunits (but exhibited no OTC enzyme activity), and one that failed to produce any detectable mitochondrial proteins. The first class of mutants divided into three complementation groups called mif 1, 2, and 3. Mif 1 and 2 respectively encode the PEP and MPP subunits of the matrix processing protease (and correspond to the mas 1 and mas 2 matants of Yaffe).

The class of mutants producing mature subunits but no activity at nonpermassive temperature was also further analyzed. First an assay was carried out that detects whether mature-sized OTC subunits have assembled into the active homotrimer by ability of the homotrimer to bind a substrate analog, δ *N*-phosphoroacetyl L-ornithine (PALO). Mature subunits in extracts from one mutant, called mif 4, passed directly through a PALO column (CHENG et al. 1989). In this mutant, assembly of an endogeous yeast mitochondrial protein, the F1BATPase subunit, was also assessed using a procedure of chloroform

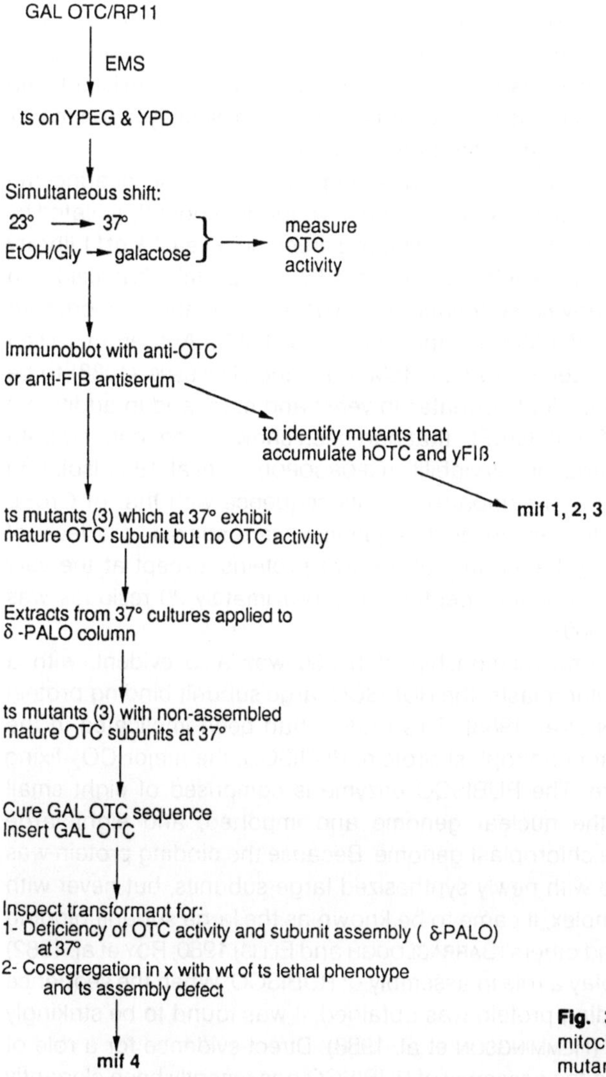

Fig. 3. Scheme for selection of mitochondrial import function (*mif*) mutants. See text for details

extraction of mitochondria (DOUGLAS et al. 1977). Following such extractions, correctly assembled F1B subunits partition into the aqueous phase. In mif 4 cells incubated at 37 °C the F1B subunits partitioned entirely into the organic phase, indicating failure of assembly. In addition to failure of assembly of OTC and F1B, two different oligomeric proteins that normally pass through the matrix compartment on their way to the intermembrane space were also affected: cytochrome b2 and the Rieske Fe/S protein (DAUM et al. 1982; GASSER et al. 1982b; HARTL et al. 1986, 1987a). These proteins reached only their intermediate-sized forms. To verify that assembly of proteins was affected, and not their translocation into the matrix space, the same proteins that had been examined in vivo were synthesized in vitro, incubated with mitochondria isolated from wild-type or mutant cells, and their localization assessed using protease protection assays. In all cases the proteins reached protected locations, consistent with their having been imported to the matrix. Thus, events of assembly in the matrix compartment appeared to be affected in the mutant.

The mif 4 phenotype was rescued by a single copy plasmid bearing the nuclear gene for a protein called hsp60. This gene was independently isolated by Hallberg and coworkers (READING et al. 1989) by screening a yeast λ gt11 library using an antiserum they had developed against a 58-kDa protein that resides in the mitochondira of *Tetrahymena, thermophila* whose abundance is increased two-fold by heat shock (MCMULLIN and HALLBERG 1987). Antiserum raised against this protein had been shown by MCMULLEN and HALLBERG (1988) to be able to recognize mitochondiral cognates in yeast and man, and in addition it recognized a protein of *E. coli*, GroEL, that had been shown to be required both for page head assembly and for cell viability (GEORGOPOULOS et al. 1983. Isolation of the yeast hsp60 gene and comparison of its sequence with that of GroEL revealed that the predicted amino acid sequences were shared to a level of approximately 50% along the lengths of the two proteins, except at the very terminus of hsp60, where a leader peptide of approximately 20 residues was present (READING et al. 1989).

An additional structural relationship of hsp60 was also evident, with a protein found in plant chloroplasts, the RUBISCO large subunit binding protein (α subunit) (HEMMINGSEN et al. 1988). This protein had been implicated in the assembly of the abundant cholroplast protein, RUBISCO, the major CO_2-fixing enzyme of the biosphere. The RUBISCO enzyme is comprised of eight small subunits, encoded by the nuclear genome and imported, and eight large subunits encoded by the chloroplast genome. Because the binding protein was first detected associated with newly synthesized large subunits, but never with the assembled L8S8 complex, it came to be known as the large subunit binding protein, and Ellis, Roy, and others (BARRACLOUGH and ELLIS) 1980; ROY et al. 1982) speculated that it must play a role in assembly of RUBISCO. When the sequence of the large subunit binding protein was obtained, it was found to be strikingly similar to that of GroEL (HEMMINGSON et al. 1988). Direct evidence for a role of GroEL-related proteins in the assembly of RUBISCO has recently been elegantly provided by LORIMER and colleagues (GOULOUBINOFF et al. 1989). They programmed *E. coli* for the synthesis of bacterial RUBISCO and demonstrated that GroEL was required in order to obtain assembly of active RUBISCO. Thus the biochemical findings of defective protein assembly in an hsp60 mutant and the structural relationship with two proteins from divergent species, both implicated in protein assembly, left one to conclude that hsp60 must play a role in protein assembly. How do these homologous, apparently endosymbiotically derived proteins work? How do they carry out the "chaperoning," as ELLIS (1988) has termed it, of other proteins?

The structure of hsp60 has been examined in cell extracts. MCMULLEN and HALLBERG (1988) have shown that hsp60 in mitochondrial matrix extracts appears in EM as a "double-donut" structure with sevenfold radial symmetry, suggesting 14-mer units of hsp60. Also, earlier studies by HENDRIX (1979) and HOHN et al. (1979) examining GroEL and PSUHKIN et al. (1982) examining chloroplasts indicate a "double donut" structure. Interestingly, mutational alteration of hsp 60 in mif 4 leads to altered behaviour of this complex—at 37 °C the complex appears to become insoluble, sedimenting into a 15000 *g* pellet. Mitochondrial subunits in the mutant behave similarly. Thus the complex appears to be required to maintain itself and mitochondrial proteins in a "soluble" state. How is this accomplished?

Recent studies indicate that hsp60 binds to unfolded polypeptides and then folds them in a process requiring ATP hydrolysis (OSTERMANN et al. 1989). When a fusion protein joining the leader peptide FI/FO ATPase subunit IX with DHFR was imported into mitochondria in the absence of ATP (by dilution from urea), the DHFR became associated with the hsp60 complex. In this association the DHFR was found to be exquisitely sensitive to proteinase K and, further, was recognizable with an antiserum against denatured DHFR. This indicates that the DHFR at the surface of hsp60 was present in an unfolded conformation. If ATP was added to the organelles the DHFR was now found in a relatively proteinase K-resistant conformation released from the hsp60 complex. This suggested that it had been folded. If the DHFR-hsp60 complex was first isolated from the matrix and ATP then added the DHFR became only partially protease-resistant and remained associated with hsp60. This indicates that folding occurs at the surface of the hsp60 complex and that probably a second factor is required to complete the reaction. Both phage head assembly and RUBISCO assembly in *E. coli* require the expression of at least one product other than the GroEL protein, the GroES protein (GOLOUBINOFF et al. 1989), a complex of 10Kd subunits. Thus it seems likely that such a component is present also in mitochondria and cooperates with hsp60.

It must be asked whether hsp60 is related structurally or functionally to another heat-inducible gene product that has recently been localized to mitochondria. This product is a member of the hsp70 gene family, but in contrast to the cytosolic family of hsp70 proteins it is encoded by a single copy nuclear gene, called *Ssc 1*, that is essential to the growth of yeast (CRAIG et al. 1987). There is no obvious structural relationship between hsp60 and the hsp 70 genes analyzed to date. Also, as yet, no functional connection between hsp60 and hsp70 has been discovered. Perhaps hsp70 in mitochondria represents a functional counterpart of the *E. coli* hsp70 homolog called dna K (BARDWELL and CRAIG 1984), which has been shown to play a role in phage DNA replication disassociating λ P protein from helicase on the template DNA (SAKAKIBARA 1988). Maybe the Ssc 1 gene product as such will prove to play a specific role in mitochondrial DNA synthesis and in maintenance of the mitochondrial genome. Alternatively, it could play a more general role like that of hsp70 in the cytosol (DESHAIES et al. 1988; CHIRICO et al. 1988) in maintaining conformations of proteins that are "loose" permitting their further biogenesis.

5 Summary

A dynamic picture of the mitochondrial protein import pathway is emerging, with conformational alteration a critical feature both preceding and following membrane translocation. The mediators of these steps of conformational alteration, as well as steps of recognition, translocation, and proteolytic cleavage, appear to be proteins. Using powerful tools of genetics and biochemistry, in years to come it should be possible to determine the precise molecular function of these proteins in mediating these novel reactions.

References

Ades I, Butow R (1980) The transport of proteins into yeast mitochondria. Kinetics and pools. J Biol Chem 255: 9925–9935

Allison Ds, Schatz G (1986) Artificial mitochondrial presequences. Proc Natl Acad Sci USA 83: 9011–9015

Amaya Y, Arakawa H, Takiguchi M, Ebina Y, Yokota S, Mori M (1988) A noncleavable signal for mitochondrial import of 3-oxoacyl-CoA thiolase. J Biol Chem 263: 14463–14470

Arakawa H, Takiguchi M, Amaya Y, Nagata S, Hargashi H, Mori M (1987) cDNA-derived amino acid sequence of rat mitochondrial 3-oxoacyl CoA thiolase with no transient presequence. EMBO J 6: 1361–1366

Argan C, Lusty CJ, Shore GC (1983) Membrane and cytosolic components affecting transport of the precursor for ornithine carbamyltransferase into mitochondria. J Biol Chem 258: 6667–6670

Audigier Y, Friedlander M, Blobel G (1987) Multiple topogenic sequences in bovine opsin. Proc Natl Acad Sci USA 84: 5783–5787

Baker A, Schatz G (1987) Sequences from a prokaryotic genome or the mouse dihydrofolate reductase gene can restore the import of a truncated precursor protein into yeast mitochondria. Proc Natl Acad Sci USA 84: 3117–3121

Bardwell JCA, Craig EA (1984) Major heat shock gene of *Drosophila* and the *Escherichia coli* heat-inducible dnaK gene are homologous. Proc Natl Acad Sci USA 81: 848–852

Barraclough R, Ellis RJ (1980) Assembly of newly synthesized large subunits into ribulose bisphosphate carboxylase in isolated pea chloroplasts. Biochim Biophys Acta 608: 19–31

Bedwell DM, Lkionsky DJ, Emr SD (1987) The yeast F_1-ATPase beta subunit precursor contains functionally redundant mitochondrial protein import information. Mol Cell Biol 7: 4038–4037

Bedwell DM, Strobel SA, Yun K, Jangeward GD, Emr S (1989) Sequence and structural requirements of a mitochondrial protein import signal defined by saturation cassette mutagenesis. Mol Cell Biol 9: 1014–1025

Berkhout TA, Rietveld A, de Kruijff B (1987) Preferential lipid association and mode of penetration of apocytochrome c in mixed model membranes as monitored by tryptophanyl fluorescence quenching using brominated phospholipids. Biochim Biophys Acta 897: 1–4

Bohni PC, Daum G, Schatz G (1983) Partial prufication of a matrix-localized protease involved in cleavage of mitochondrial precursor polypeptides. J Biol Chem 258: 4937–4943

Bohni PC, Deshaies RJ, Shekman RW (1988) SEC11 is required for signal peptide processing and yeast cell growth, J. Cell Biol 106: 1035–1042

Brandriss MC, Krzywicki KA (1986) Amino-terminal fragments of delta-pyrroline-5-carboxylate dehydrogenase direct beta-galactosidase to the mitochondrial matrix in *Saccharomyces cerevisiae*. Mol Cell Biol 6: 3502–3512

Chen W-J, Douglas M (1987a) Phosphodiester bond cleavage outside mitochondria is required for the completion of protein import into the mitochondrial matrix. Cell 49: 651–658

Chen W-J, Douglas MG (1987b) The role of protein structure in the mitochondrial import pathway—analysis of the soluble F_1-ATPase beta-subunit precursor. J Biol Chem 262: 15598–15604

Chen W-J, Douglas MG (1987c) The role of protein structure in the mitochondrial import pathway—unfolding of mitochondirally bound precursor is required for membrane translocation. J Biol Chem 262: 15605–15609

Chen W-J, Douglas MG (1988) An F_1-ATPase beta-subunit precursor lacking an internal tetramer-forming domain is imorted into mitochondria in the absence of ATP. J Biol Chem 263: 4997–5000

Cheng MY, Hartl F-U, Martin J, Pollock RA, Kalousek F, Neupert W, Hallberg EM, Hallberg RL, Horwich AL (1989) Mitochondrial heatshock protein hsp60 is essential for assembly of proteins imported into yeast mitochondria. Nature 337: 620–625

Cheng MY, Pollock RA, Hendrick JP, Horwich AL (1987) Import and processing of human ornithine transcarbamoylase precursor by mitochondria from *Saccharomyces cerevisiae.* Proc Natl Acad Sci USA 84: 4063–4067

Chirico WJ, Waters G, Blobel G (1988) 70K heat shock related proteins stimulate protein translocation into microsomes. Nature 332: 805–810

Chu TW, Grant PM, Strauss AW (1987a) Mutation of a neutral amino acid in the transit peptide of rat mitochondrial malate dehydrogenase abolishes binding and import. J Biol Chem 262: 15759–15764

Chu TW, Grant PM, Strauss AW (1987b) The role of arginine residues in the rat mitochondrial malate dehydrogenase transit peptide. J Biol Chem 262: 12806–12811

Collier DN, Bankaitis VA, Weiss TB, Bassford PJ Jr (1988) The antifolding activity of SecB promotes the export of the *E. coli* maltose binding protein. Cell 53: 273–283

Conboy J, and Rosenberg LE (1981) Posttranslational uptake and processing of in vitro synthesized ornithine transcarbamoylase precursor by isolated rat liver mitochondira. Proc Natl Acad Sci USA 78: 3073–3077

Conboy JG, Fenton WA, Rosenberg LE (1982) Processing of preornithine transcarbamylase requires a zinc-dependent protease localized to the mitochondrial matrix. Biochem Biophys Res Commun 105:1–7

Craig EA, Kramer J, Kosic-Smithers J (1987) SSC1, a member of the 70-kDa heat shock protein multigene family of *S. cerevisiae,* is essential for growth. Proc Natl Acad Sci USA 84: 4156–4160

Crooke E, Guthrie B, Stewart L, Roland L, Wickner W (1988) ProOmpA is stabilized for membrane translocation by either purified *E. coli* trigger factor or canine signal recognition particle. Cell 54: 1003–1011

Cumsky MG, Ko C, Trueblood CE, Poyton RO (1985) Two nonidentical forms of subunit V are functional in yeast cytochrome c oxidase. Proc Natl Acad Sci USA 82: 2235–2239

Cunningham K, Lill R, Crooke E, Rice M, Moore K, Wickner W, Oliver D (1989) SecA protein, a peripheral protein of the *Escherichia coli* plasma membrane, is essential for the functional binding and translocation of proOmpA. EMBO J 8: 955–959

Daum G, Gasser SM, Schatz G (1982) Import of proteins into mitochondria: energy-dependent, two-step processing of the intermembrane space enzyme cytochorme b2 by isolated yeast mitochondria. J Biol Chem 257: 13075–13080

Deshaies RJ, Koch BD, Werner-Washburne M, Craig EA, Schekman R (1988) A subfamily of stress proteins facilitates translocation of secretory and mitochondrial precursor polypeptides. Nature 332: 800–805

Douglas MG, Koy Y, Docktor ME, Schatz G (1977) Aurovertin binds to the beta subunit of yeast mitochondiral ATPase. J Biol Chem 252: 8333–8335

Douglas MG, Geller BL, Emr SD (1984) Intracellular targeting and import of an F_1-ATPase beta-subunit-beta-galactosidase hybrid protein into yeast mitochondria. Proc Natl Acad USA 81: 3983–3987

Eilers M, Schatz G (1986) Binding of a specific ligand inhibits import of a purified precursor protein into mitochondria. Nature 322: 228–232

Eilers M, Oppliger W, Schatz G (1987) Both ATP and an energized inner membrane are required to import a purified precursor protein into mitochondria. EMBO J 6: 1073–1077

Eilers M, Hwang S, Schatz G (1988) Unfolding and refolding of a purified precursor protein during import into isolated mitochondria. EMBO J 7: 1139–1145

Ellis RJ (1988) Proteins as molecular chaperones. Nature 328: 228–229

Emr SD, Vassarotti A, Garrett J, Geller BL, Takeda M, Douglas MG (1986) The amino terminus of the yeast F_1-ATPase beta-subunit precursor functions as a mitochondrial import signal. J Cell Biol 102: 523–533

Endo T, Schatz G (1988) Latent membrane perturbation activity of a mitochondrial precursor protein is exposed by unfolding. EMBO J 1153–1158

Epand RM, Hui S-W, Argan C, Gillespie LL, Shore GC (1986) Structural analysis and amphiphilic properties of a chemically synthesized mitochondrial signal peptide. J Biol Chem 22: 10017–10020

Freitag H, Janes M, Neupert W (1982) Biogenesis of mitochondrial porin and insertion into the outer mitochodrial membrane of *Neurospora crassa*. Eur J Biochem 126: 197–202

Gebellini N, Harnisch U, McCarthy JEG, Hausku G, Sebald W (1985) Clonign and expression of the fbc operon encoding the Fe/S proteins, cytochrome b and cytochrome C_1 from *Rhodopseudomonas sphaeroides* b/c_1 complex. EMBO J 4: 549–553

Gasser SM, Daum G, Schatz G (1982a) Import of proteins into mitochondria. Energy-dependent uptake of precursors by isolated mitochondria. J Biol Chem 257: 13034–13041

Gasser SM, Ohashi A, Daum G, Bohni PC, Gibson J, Reid GA, Yonetani T, Schatz G (1982b) Imported mitochondrial proteins cytochrome b2 and cytochrome c_1 are processed in two steps. Proc Natl Acad Sci USA 79: 267–271

Gavel Y, Nilsson L, von Heijne G (1988) Mitochondrial targeting sequences. Why 'non-amphiphilic' peptides may still be amphiphilic. FEBS Lett 235: 173–177

Georgopoulos CP, Tilly K, Casjens SR (1983) Lambdoid phage head assembly. In: Hendrix RW, Roberts JW, Stahl FW, Weisberg RA (eds) Lambda II. Cold Spring Harbor Laboratory, Cold Spring Harbor, pp 279–304

Gillespie LL (1987) Identification of an outer mitochondrial membrane protein that interacts with a synthetic signal peptide. J Biol Chem 262: 7939–7942

Glaser SM, Trueblood CE, Dircks LK, Poyton RD, Cumsky MG (1988) Functional analysis of mitochondrial protein import in yeast. J Cell Biol 36: 275–287

Goloubinoff P, Gatenby AA, Lorimer G (1989) GroE heat shock proteins promote assembly of foreign prokaryotic ribulose bisphosphate carboxylase oligomers in *Escherichia coli*. Nature 337: 44–47

Hackenbrock CR (1968) Ultrastructural bases for metabolically linked mechanical activity in mitochondria. II. Electron transport-linked ultrastructural transformations in mitochondria. J Cell Biol 37: 345–369

Harmey MA, Hallermayer G, Korb H, Neupert W (1977) Transport of cytoplasmically synthesized proteins into the mitochondria in a cell free system from *Neurospora crassa*. Eur J Biochem 81: 533–544

Harmey MA, Neupert W (1979) Biosynthesis of mitochondrial citrate systhase in *Neurospora crassa*. FEBS Lett 108: 385–389

Hartl F-U, Schmidt B, Wachter E, Weiss H, Neupert W (1986) Transport into mitochondria and intramitochondrial sorting of the Fe/S protein of ubiquinol-cytochrome c reductase. Cell 47: 939–951

Hartl F-U, Ostermann J, Guiard B, Neupert W (1987a) Successive translocation into and out of the mitochondrial matrix: targeting of proteins to the intermembrane space by a bipartite signal sequence. Cell 51: 1027–1037

Hartl F-U, Ostermann J, Pfanner N, Tropschug M, Guiard B, Neupert W (1987b) Import of cytochromes b2 and c_1 into mitochondria is dependent on both membrane potential and nucleoside triphosphates. In: Papa S, Chance B, Ernster L (eds) Cytochrome systems. Plenum, New York, pp 189–196

Hartl F-U, Pfanner N, Nicholson DW, Neupert W (1989) Mitochondrial protein import. Biochim Biophys Acta 988: 1–45

Hase T, Riezman H, Suda K, Schatz G (1983) Import of proteins into mitochondria: mucleotide sequence of the gene for a 70-kd protein of yeast mitochondrial outer membrane. EMBO J Biol 2: 2168–2172

Hase T, Mueller U, Riezman H, Schatz G (1984) A 70-kd protein of the yeast mitochondrial outer membrane is targeted and anchored via its extreme amino terminus. EMBO J 3: 3157–3164

Hase T, Nakai M, Matsubara H (1986) The N-terminal 21 amino acids of a 70Kd protein of the yeast mitochondrial outermembrane direct *E. coli* betagalactosidase into the mitochondrial matrix space in yeast cells. FEBS Lett 197: 199–203

Hata A, Tsuzuki T, Shimada K, Takiguch M, Mori M, Matsuda I (1988) Structure of the human ornithine transcarbamylase gene. J Biochem 103: 302–308

Hawlitschek G, Schneider H, Schmidt B, Tropschug M, Hartl F-U, Neupert W (1988) Mitochondrial protein import: identification of processing peptidase and of PEP, a processing enhancing protein Cell 53: 795–806

Hemmingsen SM, Woolford C, vanderVies SM, Tilly K, Dennis DT, Georgopoulos CP, Hendrix RW, Ellis RJ (1988) Homologous plant and bacterial proteins chaperone oligomeric protein assembly. Nature 333: 330–334

Hendrick JP, Hodges PE, Rosenberg LE (1989) A survey of amino-terminal proteolytic cleavage sites in mitochondrial precursor proteins: a three-amino-acid motif found in leader peptides that are cleaved by two matrix proteases. Proc Natl Acad Sci USA (in press)
Hendrix RW (1979) Purification and properties of groE, a host protein involved in bacteriophage assembly. J Mol Biol 129: 375–392
Hennig B, Koehler H, Neupert W (1983) Receptor sites involved in posttranslational transport of apocytochrome c into mitochondria: specificity, affinity, and number of sites. Proc Natl Acad Sci USA 80: 4963–4967
Hohn T, Hohn B, Engel A, Wurtz M (1979) Isolation and characterization of the host protein groE involved in bacteriophage lambda assembly. J Mol Biol 129: 359–373
Horwich AL, Kalousek F, Mellman I, Rosenberg LE (1985a) A leader peptide is sufficient to direct mitochondrial import of a chimeric protein. EMBO J 4: 1129–1135
Horwich Al, Kalousek F, Rosenberg LE (1985b) Arginine in the leader peptide is required for both import and proteolytic cleavage of a mitochondrial precursor. Proc Natl Acad Sci USA 82: 4930–4933
Horwich AL, Kalousek F, Fenton WA, Pollock RA, Rosenberg LE (1986) Targeting of pre-ornithine transcarbamylase to mitochondria: definition of critical regions and residues in the leader peptide. Cell 44: 451–459
Horwich AL, Kalousek F, Fenton WA, Furtak K, Pollock RA, Rosenberg LE (1987) The ornithine transcarbamylase leader peptide directs mitochondrial import through both its midportion structure and net positive charge. J Cell Biol 105: 669–677
Hsu YP, Weyler W, Chen S, Sims KS, Rinehart WB, Utterback MC, Powell JF, Breakefield XO (1988) Structural features of human monoamine oxidase A elucidated from cDNA and peptide sequences. J Neurochem 51: 1321–1324
Hurt EC, Pesold-Hurt B, Schatz G (1984) The cleavable prepiece of an imported mitochondrial protein is sufficient to direct cytosolic dihydrofolate reductase into the mitochondrial matrix. FEBS Lett 178: 306–310
Hurt EC, Schatz G (1987) A cytosolic protein contains a cryptic mitochondrial targeting signal. Nature 325: 499–503
Hurt EC, Pesold-Hurt B, Suda K, Oppliger W, Schatz S (1985) The first twelve amino acids (less than half of the pre-sequence) of an imported mitochondrial protein can direct mouse cytosolic dihydrofolate reductase into the yeast mitochondrial matrix. EMBO J 4: 2061–2068
Hurt EC, Goldschmidt-Clermont M, Pesold-Hurt B, Rochaix JD, Schatz G (1986) A mitochondrial presequence can transport a chloroplast-encoded protein into yeast mitochondria. J Biol Chem 261: 11440–114413
Hurt EC, Allison DA, Schatz G (1987) Amino-terminal deletions in the presequence of an imported mitochondrial protein block the targeting function and proteolytic cleavage of the presequence at the carboxy terminus. J Biol Chem 262: 1420–1424
Huygen R, Crabeel M, Glansdorf N (1987) Nucleotide sequnce of the ARG3 gene of the yeast *Saccharomyces cerevisiae* encoding ornithine carbamoyltransferase. Comparison with other carbamoyltransferases. Eur J Bio Chem 166: 371–377
Isaya G, Fenton WA, Hendrick JP, Furtak K, Kalousek F, Rosenberg LE (1988) Mitochondrial import and processing of mutant human ornithine transcarbamylase precursors in cultured cells. Mol Cell Biol 8: 5150–5158
Ito K, Wittekind M, Nomura M, Shiba K, Yura T, Nashimoto H (1983) A temperature-sensitive mutant of *E. coli* exhibiting slow processing of exported proteins. Cell 32: 789–797
Jensen RE, Yaffe MP (1988) Import of proteins into yeast mitochondria: the nuclear MAS2 gene encodes a component of the processing protease that is homologous to the MAS1-encoded subunit. EMBO J 7: 3863–3871
Kalousek F, Francois B, Rosenberg LE (1978) Isolation and characterization of ornithine transcarbamylase from normal human liver. J Biol Chem 253: 3939
Kalousek F, Orsulak MD, Rosenberg LE (1984) Newly processed ornithine transcarbamylase subunits are assembled to trimers in rat liver mitochondria. J Biol Chem 259: 5392–5395
Kalousek F, Hendrick JP, Rosenberg LE (1988) Two mitochondrial matrix proteases act sequentially in the processing of mammalian matrix enzymes. Proc Natl Acad Sci USA 85: 7536–7540
Kellems RE, Allison VF, Butow RA (1974) Cytoplasmic type 80S ribosomes associated with yeast mitochondria II. Evidence for the association of cytoplasmic ribosomes with the outer mitochondrial membrane in situ. J Biol Chem 249: 3297–3303
Kellems RE, Allison VF, Butow RA (1975) Cytoplasmic type 80S ribosomes associated with yeast

mitochondria IV. Attachment of ribosomes ot the outer membrane of isolated mitochondria. J Cell Biol 65: 1–14

Keng T, Alani E, Guarente L (1986) The nine amino-terminal residues of delta-aminolevulinate synthase direct beta-galactosidase into the mitochondrial matrix. Mol Cell Biol 6: 355–364

Kleene R, Pfanner N, Pfaller R, Link TA, Sebald W, Neupert W, Tropschug M (1987) Mitochondrial porin of *Neurospora crassa:* cDNA cloning, in vitro expression and import into mitochondria. EMBO J 6: 2627–2634

Kolansky DM, Conboy JG, Fenton WA, Rosenberg L (1982) Energy-dependent translocation of the precursor of ornithine transcarbamylase by isolated rat liver mitochondria. J Biol Chem 257: 8467-8471

Korb H, Neupert W (1978) Biogenesis of cytochrome c in *Neurospora crassa:* synthesis of apocytochrome c, transfer to mitochondria, and conversion to holocytochome c. Eur J Biochem 91: 619–620

Kraus JP, Novotny J, Kalousek F, Swaroop M, Rosenberg LE (1988) Different structures in the amino-terminal domain of the ornithine transcarbamylase leader peptide are involved in mitochondrial import and corboxyl-terminal cleavage. Proc Natl Acad Sci USA 85: 8905–8909

Lagace M, Howell BW, Burak R, Lusty CJ, Shore GC (1987) Rat carbamyl-phosphate synthetase I gene. Promoter sequence and tissue-specific transcriptional regulation in vitro. J Biol Chem 262: 10415–10418

Lill R, Crooke E. Guthrie B, Wickner W (1988) The "trigger factor cycle" includes ribosomes? presecretory proteins, and the plasma membrane. Cell 54: 1013–1018

Lill R, Cunningham K, Brundage LA, Ito K, Oliver D, Wickner W (1989) SecA protein hydrolyzes ATP and is an essential component of the protein translocation ATPase of *Escherichia coli.* EMBO J 8: 961–966

Liu XQ, Bell AW, Freeman KB, Shore GC (1988) Topogenesis of mitochondrial inner membrane uncoupling protein. Rerouting transmembrane segments to the soluble matrix compartment. J Cell Biol 107: 503–509

Maarse AC, van Loon AP, Riezman H, Gregor I, Schatz G, Gravell LA (1984) Subunit IV of yeast cytochrome c oxidase: cloning and nucleotide sequencing of the mature protein. EMBO J 3: 2831–2837

Maccecchini ML, Rudin Y, Schatz G (1979a) Transport of proteins across the mitochondrial outer membrane: a precursor form of the cytoplasmically made intermembrane enzyme cytochrome c peroxidase. J Biol Chem 254: 7468–7471

Maccecchini ML, Rudin Y, Blobel G, Schatz G (1979b) Import of proteins into mitochondria: precursor forms of the extramitochondrially made F_1 ATPase subunits in yeast. Proc Natl Acad Sci USA 76: 343–347

Maguire DJ, Day AR, Borthwich IA, Srivastava G, Wigley PL, May BK, Elliot WH (1986) Nucleotide sequence of the chicken 5-aminolevulinate synthase gene. Nucl Acids Res 14: 1379–1391

McAda PC, Douglas M (1982) A neutral metalloendoprotease involved in the processing of an F_1-ATPase subunit precursor in mitochondria. J Biol Chem 257: 3177–3182

McMullin TW, Hallberg RL (1987) A normal mitochondrial protein is slectively synthesized and accumulated during heat shock in *Tetrahymena thermophila.* Mol Cell Biol 7: 4414–4423

McMullin TW, Hallberg Rl (1988) A highly evolutionarily conserved mitochondrial protein is structurally related to the protein encoded by the *E. coli* groEL gene. Mol Cell Biol 8: 371–380

Michel R, Wachter E, Sebald W (1979) Synthesis of a larger precursor for the proteolipid subunit of the mitochondrial ATPase complex of *Neurospora crassa* in a cell-free wheat germ system. FEBS Lett 101: 373–376

Mihara K, Sato R (1985) Molecular cloning and sequencing of cDNA for yeast protein, an outer mitochondrial membrane protein: a search for targeting signal in the primary structure. EMBO J 4: 769–774

Miura S, Mori M, Amaya Y, Tatibana M (1982) A mitochondrial protease that cleaves the precursor of ornithine carbamyltransferase. Eur J Biochem 122: 641–647

Mori M, Matsue H, Miura S, Tatibana M, Hashimoto T (1985) Transport of proteins into mitochondrial matrix. Evidence suggesting a common pathway for 3-ketoacyl-CoA thiolase and enzymes having presequences. Eur J Biochem 149: 181–186

Murakami H, Pain D, Blobel G (1988) 70-kD heat shock-related protein is one of at least two distinct cytosolic factors that stimulate protein import into mitochondria. J Cell Biol 107: 2051–2067

Murakami K, Amaya Y, Takiguchi M, Ebina Y, Mori M (1988) Reconstitution of mitochondrial protein transport with purified ornithine carbamoyltransferase precursor expressed in *Escherichia coli.* J Biol Chem 263: 18437–18442

Nargang FE, Drygas ME, Kwong PL, Nicholson DW, Neupert W (1988) a mutant of *Neurospora crassa* deficient in cytochrome c heme lyase activity cannot import cytochrome c into mitochondria. J Biol Chem 263: 9388–9394

Nguyen M, Argan C, Lusty CJ, Shore GC (1986) Import and processing of hybrid proteins by mammalian mitochondria in vitro. J Biol Chem 261: 800–805

Nguyen M, Argan C, Sheffield WP, Bell AW, Shields D, Shore GC (1987) A signal sequence domain essential for processing, but not import, of mitochondrial pre-ornithine carbamyltransferase. J Biol Chem 104: 1193–1198

Nguyen M, Bell AW, Shore GC (1988) Protein sorting between mitochondrial membranes specified by position of the stop-transfer domain. J Cell Biol 106: 1499–1505

Nicholson DW, Neupert W (1990) Import of cytochrome c into mitochondria: reduction of heme, mediated by NADH and flavin nucleotides, is obligatory for its covalent linkage to apocytochrome c. Proc Natl Acad Sci USA

Nicholson DW, Kohler H, Neupert W (1987) Import of cytochrome c into mitochondria. Cytochrome c heme lyase. Eur J Biochem 164: 147–157

Nicholson DW, Hergersberg C, Neupert W (1988) Role of cytochrome c heme lyase in the import of cytochrome c into mitochondria. J Biol Chem 263: 19034–19042

Ohashi A, Gibson J, Gregor I, Schatz G (1982) Import of proteins into mitochondria: the precursor of cytochrome c_1 is processed in two steps, one of them heme-dependent. J Biol Chem 257: 13042–13047

Ohba M, Schatz G (1987a) Protein import into yeast mitochondria is inhibited by antibodies raised against 45-kd proteins of the outer membrane. EMBO J 6: 2109–2115

Ohba M, Schatz G (1987b) Disruption of the outer membrane restores protein import to trypsin-treated yeast mitochondria. EMBO J 6: 2117–2122

Ohta S, Schatz G (1984) A purified precursor polypeptide requires a cytosolic protein fraction for import into mitochondria. EMBO J 3: 651–657

Ono H, Tuboi S (1988) The cytosolic factor required for import of precursors of mitochondrial proteins into mitochondria. J Biol Chem 263: 3188-3193

Ostermann J, Horwich AL, Neupert W, Hartl F-U (1989) Protein folding in mitochondria requires complex formation with hsp60 and ATP hydrolysis. Nature 341: 125–130

Pain D, Kanwar YS, Blobel G (1988) Identification of a receptor for protein import into chloroplasts and its localization to envelope contact zones. Nature 331: 232–237

Pelham HR (1986) Speculations of the functions of the major heat shock and glucose-regulated proteins. Cell 46: 959–961

Pfaller R, Neupert W (1987) High-affinity binding sites involved in the import of porin into mitochondria. EMBO J 6: 2635–2642

Pfaller R, Freitag H, Harmey MA, Benz R, Neupert W (1985) A water-soluble form of porin from the mitochondrial outer membrane of *Neurospora crassa*—properties and relationship to the biosynthetic precursor form. J Biol Chem 260: 8188–8193

Pfaller R, Steger HF, Rassow J, Pfanner N, Neupert W (1988) Import pathways of precursor proteins into mitochondria: multiple receptor sites are followed by a common membrane insertion site. J Cell Biol 107: 2483–2490

Pfanner N, Neupert W (1985) Transport of proteins into mitochondria: a potassium diffusion potential is able to drive the import of ADP/ATP carrier. EMBO J 4: 2819–2825

Pfanner N, Neupert W (1986) Transport of F1-ARPase subunit beta into mitochondria depends on both a membrane potential and nucleoside triphosphates. FEBS Lett 209: 152–156

Pfanner N, Neupert W (1987) Distinct steps in the import of ADP/ATP carrier into mitochondria. J Biol Chem 262: 7528–7536

Pfanner N, Hartl H-U, Guiard B, Neupert W (1987a) Mitochondrial precursor proteins are imported through a hydrophilic membrane environment. Eur J Biochem 169; 289–293

Pfanner N, Hoeben P, Tropschug M, Neupert W (1987b) The carboxyl-terminal two thirds of the ADP/ATP carrier polypeptide contains sufficient information to direct translocation into mitochondria. J Biol Chem 262: 14851–14854

Pfanner N, Muller HK, Harmey MA, Neupert W (1987c) Mitochondrial protein import: involvement of the mature part of a cleavable precursor protein in the binding to receptor sites. EMBO J 6: 3449–3454

Pfanner N, Tropschug M, Neupert W (1987d) Mitochondrial protein import: nucleotide triphosphates are involved in conferring import competence to precursors. Cell 49: 815–823

Pfanner N, Pfaller R, Kleene R, Ito M, Tropschug M, Neupert W (1988) Role of ATP in mitochondrial protein import. Conformational alteration of a precursor protein can substitute for ATP requirement. J Biol Chem 263: 4049–4051

Pilgrim D, Young ET (1987) Primary structure requirements for correct sorting fo the yeast mitochondrial protein ADH III to the yeast mitochondrial matrix space. Mol Cell Biol 7: 294–304

Pollock RA, Hartl F-U, Cheng MY, Ostemann J, Horwich A, Neupert W (1988) The processing peptidase of yeast mitochondria: the two cooperating components MPP and PEP are structurally related. EMBO J 7:3493–3500

Pratje E, Guiard B (1986) One nuclear gene controls the removal of transient pre-sequences from two yeast proteins: one encoded by the nuclear the other by the mitochondrial genome. EMBO j 5: 1313–1317

Pratje E, Mannhaupt G, Michaelis G, Beyreuther K (1983) A nuclear mutation prevents processing of a mitochondrially encoded membrane protein in *Saccharomyces cerevisiae*. EMBO J 2: 1049–1054

Puskhin AV, Tsuprun VL, Solovjeva NA, Shubin VV, Erstigneeva ZG, Kretovich WL (1982) High molecular weight pea leaf protein similar to the GroE protein of *Escherichia coli*. Biochim Biophys Acta 704: 379–384

Rassow J, Guiard B, Wienhues U, Herzog V, Hartl F-U, Neupert W (1990) Translocation arrest by reversible folding of a precursor protein imported into mitochondria. A means to quantitate translocation contact sites. J Cell Biol

Reading DS, Hallberg RL, Myers AM (1989) Characterization of the yeast hsp60 gene encoding a mitochondrial assembly factor. Nature 337: 655–659

Rietveld A, Sijens P, Verkliej AJ, de Kruijff B (1983) Interactions of cytochrome c and its precursor apocytochrome c with various phospholipids. EMBO J 2: 907–913

Riezman H, Hay R, Witte C, Nelosn N, Schatz G (1983) Yeast mitochondrial outer membrane specifically binds cytoplasmically synthesized precursors of mitochondrial proteins. EMBO J 2: 1113–1118

Roise D, Horvath SJ, Tomich JM, Richards JH, Schatz G (1986) A chemically synthesized pre-sequence of an imported mitochondrial protein can form an amphiphilic helix and perturb an artificial phospholipid bilayer. EMBO J 5: 1327–1334

Roy H, Bloom M, Milos P, Monroe M (1982) Studies on the assembly of large subunits of ribulose bisphosphate carboxylase in isolated pea chloroplasts. J Cell Biol 94: 20–27

Sakakibara Y (1988) The dnaK gene of *Escherichia coli* functions in initation of chromosome replication.J Bacteriol 170: 972–929

Scarpulla RC, Nye SH (1983) Functional expression of rat cytochrome c in *Saccharomyces cerevisiae*. Proc Natl Acad Sci USa 83: 6352–6356

schleyer M, Neupert W (1985) Transport of proteins into mitochondria: translocational intermediates spanning contact sites between outer and inner membranes. Cell 43: 339–350

Schleyer M, Schmidt B, Neupert W (1982) Requirement of a membrane potential for the post-translational transfer of proteins into mitochondria. Eur J Biochem 125: 109–116

Schmidt B, Hennig B, Kohler H, Neupert W (1983) Transport of the precursor to *Neurospora* ATPase subunit 9 into yeast mitochondria. J Biol Chem 258: 4687–4689

Schmidt B, Wachter E, Sebald W, Neupert W (1984) Processing peptidase of *Neurospora* mitochondria. Two step cleavage of imported ATPase subunit 9. Eur J Biochem 144: 581–588

Schwaiger M, Herzog V, Neupert W (1987) Characterization of translocation contact sites involved in the import of mitochondrial proteins. J Cell Biol 105: 235–246

Shih MC, Heinrich P, Goodman H (1988) Intron existence predated the divergence of eukaryotes and prokaryotes. Science 242: 1164–1166

Schulte U, Arretz M, Schneider H, Tropschug H, Wachter E, Neupert W, Weiss H (1989) A family of mitochondrial proteins involved in bioenergetics and biogenesis. Nature 339: 147–149

Smagula C, Douglas MG (1988) Mitochondrial import of the ADP/ATP carrier protein in *Saccharomyces cerevisiae*: sequences required for receptor binding and membrane transloca-tion. J Biol Chem 263: 6783–6790

Sollner T, Pfanner N, Neupert W (1988) Mitochondrial protein import: differential recognition of various transport intermediates by antibodies. FEBS Lett 229: 25–29

Sollner T Griffiths G, Pfaller R, Pfanner N, Neupert W (1989) MOM 19 an import receptor for mitochondrial precursor proteins. Cell 59: 1061–1070

Steiner DF, Quinn PS, Chan SJ, March J, Tager H (1980) Processing mechanisms in the biosynthesis of proteins. Ann NY Acad Sci 343: 1–16

Stuart RA, Nicholson DW, Neupert W (1990) Early steps in the import of precursor proteins into

mitochondria: receptor functions can be substituted by membrane inssertion activity of apocytochrome c (to be published)
Suissa M, Suda K, Schatz G (1984) Isolation of the nuclear yeast genes for citrate synthase and fifteen other mitochondrial proteins by a new screening method. EMBO J 3: 1773–1781
Sztul ES, Hendrick JP, Kraus JP, Wall D, Kalousek F, Rosenberg LE (1987) Import of rat ornithine transcarbamylase precursor into mitochondria: two-steps processing of the leader peptide. J Cell Biol 105: 2631–2639
Sztul ES, Chu TW, Strauss AW, Rosenberg LE (1988) Import of the malate dehydrogenase precursor by mitochondria. Cleavage within leader peptide by matrix protease leads to formation of intermediate-size form. J Biol Chem 263: 12085–12091
Takiguchi M, Miura S, Mori M, Tatibana M (1983) Transport of proteins into mitochondria: a high conservation of precursor uptake and processing system. Comp Biochem Physiol [b] 75B: 227–231
Takiguchi M, Mruakami T, Miura S Mori M (1987) Structure of the rat ornithine carbamoyltransferase gene, a large X chromosomelinked gene with an atypical promoter. Proc Natl Acad Sci USA 84: 6136–6140
Teintze M, Slaughter M, Weiss H, Neupert W (1982) Biogenesis of mitochondrial ubiquinol-cytochrome c reductase and cytochrome bc_1 complex precursor proteins and their transfer into mitochondria. J Biol Chem 257: 10364–10371
Tropshug M, Nicholson DW, Hartl F-U, Kohler H, Pfanner N, Wachter E, Neupert W (1988) Cyclosporin A-binding (cyclophilin) of *Neuspora crassa*. One gene codes for both cytosolic and mitochondrial forms. J Biol Chem 263: 14433–14440
Tzagoloff A, Mian W, Crivellone M (1986) Assembly of the membrane system. Characterization of *CorI*, the structural gene for the 44-kilodalton core protein of yeast coenzyme QH_2-cytochrome c reductase. J Biol Chem 261: 17163–17169
Urrestarazu S, Vossers S, Wiame JM (1977) Change in location of ornithine carbamoyltransferase and carbamoylphosphate synthetase among yeast in relation to the arginase/ornithine carbamoyltransferase regulation. Eur J Biochem 79: 473–481
van Loon, Schatz G (1987) Transport of proteins to the mitochondrial intermembrane space: the 'sorting' domain of the cytochrome c_1 presequence is a stop-transfer sequence specific for the mitochondrial inner membrane. EMBO J 6: 2441–2448
van Loon AP, Brandli AW, Schatz G (1986) The presequences of two imported mitochondrial proteins contain information for intracellular and intramitochondrial sorting. Cell 44: 801–812
Verner K, Schatz G (1987) Import of an incompletely folded precursor protein into isolated mitochondria requires and energized inner membrane, but no added ATP. EMBO J 6: 2449–2456
Verner K, Weber M (1989) Protein import into mitochondria in a homologous yeast in vitro system. Biol Chem 264: 3877–3879
Vestweber D, Schatz G (1988a) Mitochondria can import artificial precursor proteins containing a branched polypeptide chain or a carbosy-terminal stilbene disulfonate. J Cell Biol 107: 2045–2049
Vestweber D, Schatz G (1988b) Point mutations destabilizing a precursor protein enhance its post-translational import into mitochondria. EMBO J 7: 1147–1151
Vestweber D, Schatz G (1988c) A chimeric mitochondrial precursor protein with internal disulfide bridges blocks import of authentic precursors into mitochondria and allows quantitation of import sites. J Cell Biol 107: 2037–2043
Vestweber D, Brunner J, Baker A, Schatz G (1989) A 42k outer membrane protein is a component of the yeast mitochondrial protein import site. Nature 341: 205–209
von Heijne G (1986a) A new method for predicting signal sequence cleavage sites. Nucleic Acids Res 14: 4683–4690
von Heijne G (1986b) Mitochondrial targeting sequences may form amphiphilic helices. EMBO J 5: 1335–1342
von Heijne G (1986c) Why mitochondria need a genome. FEBS Lett 198: 1–4
von Heijne G, Steppuhn J, Herrmann RG (1989) Domain structure of mitochondrial and chloroplast targeting peptides. Eur J Biochem 180: 535–545
Walter P and Lingappa VR (1986) Mechanism of protein translocation across the endoplasmic reticulum membrane. Annu Rev Cell Biol 2:499–516
Werner-Washburne M, Stone DE, Craig E (1987) Complex interactions among members of an essential subfamily of hsp70 genes in *Saccharomyces cerevisiae*. J Cell Biol 7:2568–2577
Witte C, Jensen RE, Yaffe M, Schatz G (1988) MAS1, a gene essential for yeast mitochondrial assembly, encodes a subunit of the mitochondrial processing protease. EMBO J 7: 1439–1447

Yaffe M, Schatz G (1984) Two nuclear mutations that block mitochondrial protein import in yeast. Proc Natl Acad Sci USA 81: 4819–4823

Yaffe M, Ohta S, Schatz G (1985) A yeast mutant temperature sensitive for mitochondrial assembly is deficient in a mitochondrial protease activity that cleaves imported precursor polypeptides. EMBO J 4: 2069–2074

Yang M, Jensen RE, Yaffe MP, Oppliger W, Schatz G (1988) Import of proteins into yeast mitochondria: the purified matrix processing protease contains two subunits which are encoded by the nuclear MAS1 and MAS 2 genes. EMBO J 7: 3857–3862

Zimmermann R and Neupert W (1980) Transport of proteins into mitochondria: posttranslational transport of ADP/ATP carrier into mitochondria in vitro. Eur J Biochem 109: 217–229

Zimmermann R, Henning B, Neupert W (1981) Different transport pathways of individual precursor proteins in mitochondria. Eur J Biol Chem 116: 455–460

Zwizinski C, Neupert W (1983) Precursor proteins are transported into mitochondria in the absence of proteolytic cleavage of the additional sequences. J Biol Chem 258: 13340–13346

Zwizinski C, Schleyer M, Neupert W (1983) Transfer of proteins into mitochondria. Precursor to the ADP/ATP carrier binds to receptor sites on isolated mitochondria. J Biol Chem 258: 4071–4074

Zwizinski C, Schleyer M, Neupert W (1984) Proteinaceous receptors for the import of mitochondrial precursor proteins. J Biol Chem 259: 7850–7856

Targeting of Proteins to the Lysosome

S. R. PFEFFER

1 Introduction

Recent work from a number of laboratories is beginning to offer important clues to the mechanisms by which proteins are targeted to lysosomes. Soluble lysosomal enzymes undergo a unique posttranslational modification that facilitates their intracellular routing: mannose 6-phosphate (man6P) is attached to their asparagine-linked oligosaccharides. Intracellular receptors recognize and bind to man6P-bearing proteins, and the receptor-ligand complexes are then transported to a prelysosomal organelle. Despite the identification of man6P-specific receptors for soluble lysosomal enzymes, it is still unclear how the cell selects this class of receptors for transport to prelysosomes, rather than

Department of Biochemistry, Stanford University School of Medicine, Stanford, CA 94305-5307, USA delivering them to the cell surface or another intracellular compartment. Our lack of understanding of this sorting process is further underscored by the recent finding that lysosomal *membrane* glycoproteins never acquire man6P, yet they are nevertheless accurately targeted to lysosomes. This chapter will review our current picture of lysosomal protein targeting. Several excellent related reviews are also recommended to the reader (VON FIGURA and HASILIK 1986; KORNFELD 1986, 1987; SAHAGIAN 1987; KORNFELD and MELLMAN 1989).

2 Biosynthetic Protein Transport

Lysosomal proteins employ the secretory pathway to accomplish their delivery to lysosomes. Like proteins destined for the cell surface or secretory storage granules, lysosomal proteins contain N-terminal signal sequences which direct their co-translational translocation into the lumen of the endoplasmic reticulum (ER). Within the ER, they receive high-mannose, asparagine-linked oligosaccharides, disulfide bridges are formed, the proteins fold, and, where applicable, they assemble into their correct oligomeric structures (ROSE and DOMS 1988). They are then transported in membrane-bound vesicles to the Golgi complex, along with proteins destined for the cell surface, secretory storage granules, or the Golgi complex itself. Here, the oligosaccharide chains of most glycoproteins are trimmed and remodeled (KORNFELD and KORNFELD 1985), and it is within the Golgi complex that proteins are somehow sorted to their appropriate final destinations.

Beginning just prior to their arrival in the Golgi, soluble lysosomal enzymes are acted upon by the enzyme, *N*-acetylglucosamine phosphotransferase, which adds *N*-acetylglucosamine phosphate (P-GlcNAc) to mannose residues present on their asparagine-linked oligosaccharides (Fig. 1; VON FIGURA and HASILIK 1986; LAZZARINO and GABEL 1988). Within the Golgi complex GlcNAc is then removed, and in this manner lysosomal hydrolases acquire a man6P moiety. The hydrolases then proceed through the remainder of the Golgi complex (from cis to medial to trans Golgi), and it is not until the enzymes reach the last compartment of the Golgi complex, the trans-Golgi network (TGN; GRIFFITHS and SIMONS 1986), that they are segregated into distinct transport vesicles and away from other cell surface and secreted proteins (DUNCAN and KORNFELD 1988).

Receptors specific for phosphomannosyl-containing oligosaccharides are present in the Golgi complex, and the formation of a lysosomal enzyme-man6P receptor complex is the first step by which lysosomal enzymes are segregated from proteins with other destinations in the secretory pathway. Two distinct man6P receptors have been identified. They are referred to as "cation-dependent" and "cation-independent" man6P receptors, since one requires the presence of divalent cations for ligand binding in vitro. A detailed description of these receptors is presented below.

3 Lysosomal Enzyme Phosphorylation

KAPLAN and coworkers (1977) were the first to demonstrate that a hexose phosphate, specifically man6P, is essential for the high-affinity and saturable endocytosis of lysosomal hydrolases by cultured cells. It is now clear that over 50 different soluble lysosomal enzymes acquire man6P residues as they traverse the secretory pathway. The importance of man6P in lysosomal enzyme targeting is demonstrated rather dramatically in I-cell disease (see VON FIGURA and HASILIK 1986 for review). Fibroblasts from patients with I-cell disease (mucolipidosis II) and pseudo-Hurler polydystrophy (mucolipidosis III) are deficient in *N*-acetylglucosamine phosphotransferase activity, and thus cannot synthesize the man6P recognition marker. The absence of man6P leads to the secretion of most lysosomal enzymes from these cells.

Structural analyses have shown that five of the nine mannose residues that comprise the high-mannose oligosaccharide chains of lysosomal enzymes are potential phosphorylation sites (VARKI and KORNFELD 1980). Furthermore, a single high-mannose oligosaccharide may acquire two man6P residues (VARKI and KORNFELD 1983); such diphosphorylated oligosaccharides are the preferential ligand for both cation-dependent and cation-independent man6P receptors (CREEK and SLY 1982; FISCHER et al. 1982; VARKI and KORNFELD 1983; HOFLACK et al. 1987). The first P-GlcNAc is probably added to the α1,6 branch of the high-mannose oligosaccharide chain (Fig. 1; GOLDBERG and KORNFELD 1981; LAZZARINO and GABEL 1988). Also, an individual lysosomal enzyme is likely to contain multiple oligosaccharide chains, any of which may become phosphorylated.

N-acetylglucosamine phosphotransferase is responsible for the selectivity of phosphomannosyl addition, and therefore the selectivity of lysosomal enzyme sorting. Kornfeld and colleagues have shown that phosphotransferase recognizes a feature of the three-dimensional folded surface of lysosomal enzymes, since heat denaturation or protease fragmentation of lysosomal enzymes yields poor phosphorylation substrates (REITMAN and KORNFELD 1981; LANG et al. 1984). Importantly, lysosomal proteins are better substrates than nonlysosomal proteins by a factor of at least 100 (REITMAN and KORNFELD 1981).

Fibroblasts from certain patients with pseudo-Hurler polydystrophy possess a phosphotransferase with normal activity using α-methylmannoside as an acceptor, but with significantly decreased affinity for lysosomal hydrolases (VARKI et al. 1981; LANG et al. 1985). This has been taken to suggest that phosphotransferase has both a protein recognition site that is selective for lysosomal hydrolases and a catalytic site that binds high-mannose oligosaccharide chains (KORNFELD 1987).

A few recent, unanticipated discoveries have revealed new subtleties with regard to phosphotransferase specificity. The precursor for transforming growth factor β (TGF-β), when expressed in Chinese hamster ovary cells, and proliferin, a prolactin-related, secreted placental growth factor, both bear man6P residues and bind to man6P receptors in vitro (PURCHIO et al. 1988; LEE and NATHANS 1988). Perhaps even more surprising was the finding that the epidermal growth factor (EGF) receptor, a transmembrane glycoprotein, also bears man6P-containing oligosaccharide side chains (TODDERUD and CARPENTER 1988). It is entirely unclear why these nonlysosomal proteins should bear man6P residues, and the simplest explanation for the presence of man6P on these proteins is that it reflects a low level of nonselective phosphorylation. To better understand the consequences of this phosphorylation, it will be important to determine the fraction of the phosphorylated secretory products that are actually retained within cells.

Variability in phosphorylation has also been observed for a single protein when expressed in different cell types. FAUST et al. (1987) expressed the secretory protein, renin, in XENOPUS oocytes and in mouse L-2234 cells. In the oocytes 50%–90% of the renin became phosphorylated and remained intracellular, yet in the L-2234 cells greater than 90% of the renin was secreted and only 5%–6% contained the man6P recognition marker. Renin is related to the lysosomal protease, cathepsin, and apparently contains a remnant of the signal for specific phosphorylation. Differential phosphorylation could be explained if renin were a better substrate for the XENOPUS phosphotransferase than for the murine enzyme. Alternatively, since proteins traverse the XENOPUS secretory pathway at an extremely slow rate, it is possible that the observed phosphorylation was simply a consequence of longer exposure of this normally feeble substrate to the phosphotransferase enzyme.

It is important to note that the mere presence of man6P does not guarantee efficient targeting to lysosomes; phosphomannosyl-containing proteins are often secreted from cells. As mentioned earlier, oligosaccharides bearing two man6P groups bind with highest affinity to man6P receptors, and the weaker affinity of monophosphorylated oligosaccharides can be compensated for by the presence of two monophosphorylated side chains on a single lysosomal enzyme (cf. HOFLACK et al. 1987). If man6P receptors are limiting within the secretory pathway, diphosphorylated or multiply monophosphorylated lysosomal enzymes will bind preferentially to available receptors, resulting in efficient sorting to the lysosome. However, the orientation of even a diphosphorylated oligosaccharide on the surface of the lysosomal enzyme may render it inaccessible for man6P receptor binding (HOFLACK et al. 1987). Thus, the presence of man6P on the TGF-β precursor, proliferin, and the EGF receptor may have little consequnce for these proteins, and the significance of these modifications will require further investigation.

Phosphorylation of lysosomal enzymes is likely to occur in two stages in vivo. LAZZARINO and GABEL (1988) examined the phosphorylation of lysosomal enzymes in the presence of energy poison or after low-temperature incubation (15 °C), conditions in which protein export from the ER is blocked. These workers showed that lysosomal enzymes encountered phosphotransferase and acquired a single P-GlcNAc moiety when retarded within the ER. This was unexpected, because it was previously believed that phosphotransferase resided in the Golgi complex. Upon reversal of the export block, hydrolase oligosaccharides acquired a second P-GlcNAc, and the man6P groups were uncovered by the action of the P-GlcNAc-specific phosphoglycosidase (see Fig. 1). These data suggest that at least some of the lysosomal enzyme-specific phosphotransferase is located in a *pre-Golgi* compartment, as well as within the Golgi complex, while the phosphoglycosidase resides exclusively in the Golgi complex. These results are not inconsistent with the finding that phosphotransferase and phosphoglycosidase enzymes occupy cellular compartments that are separable by density gradient centrifugation (GOLDBERG and KORNFELD 1983; DEUTSCHER et al. 1983). In addition, this compartmentalization model readily explains the observation by PELHAM (1988) that the oligosaccharide side chains of the lysosomal enzyme, cathepsin D, acquired P-GlcNAc when this protein was artificially retained in the ER by an attached Lys-Asp-Glu-Leu ER retention signal.

An alternative explanation for two-step phosphorylation is that it reflects the action of two distinct phosphotransferases, housed in the late ER and the early Golgi complex respectively. Resolution of this issue awaits purification and further characterization of the phosphotransferase enzyme(s).

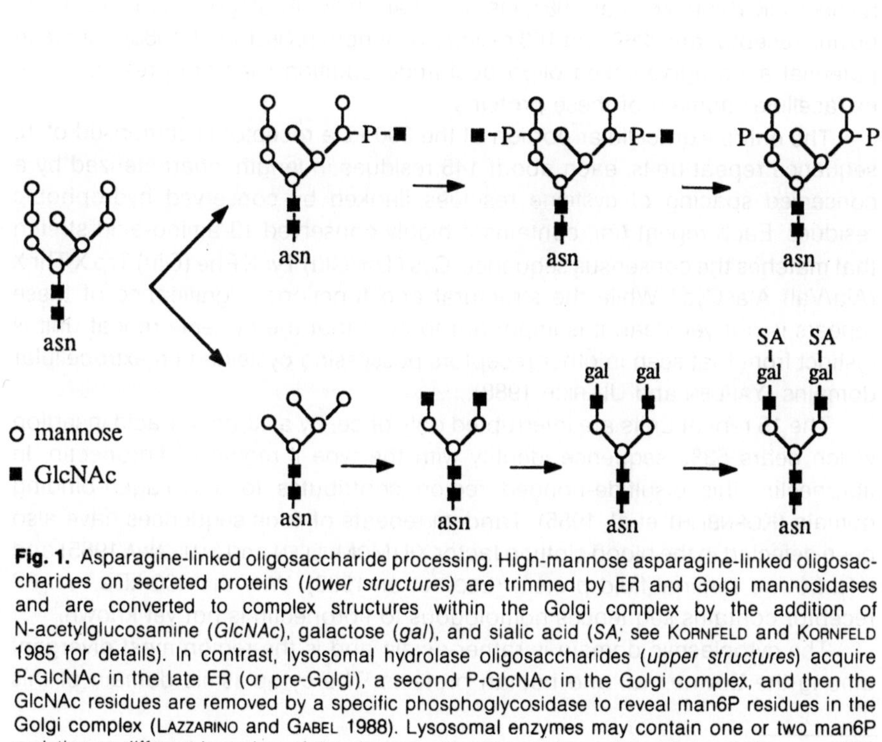

Fig. 1. Asparagine-linked oligosaccharide processing. High-mannose asparagine-linked oligosaccharides on secreted proteins (*lower structures*) are trimmed by ER and Golgi mannosidases and are converted to complex structures within the Golgi complex by the addition of N-acetylglucosamine (*GlcNAc*), galactose (*gal*), and sialic acid (*SA*; see KORNFELD and KORNFELD 1985 for details). In contrast, lysosomal hydrolase oligosaccharides (*upper structures*) acquire P-GlcNAc in the late ER (or pre-Golgi), a second P-GlcNAc in the Golgi complex, and then the GlcNAc residues are removed by a specific phosphoglycosidase to reveal man6P residues in the Golgi complex (LAZZARINO and GABEL 1988). Lysosomal enzymes may contain one or two man6P moieties on different branches of the oligosaharide. (Adapted from GOLDBERG and KORNFELD 1983; LAZZARINO and GABEL 1988)

4 The Cation-Independent Mannose 6-Phosphate Receptor

The cation-independent man6P receptor is a large transmembrane glycoprotein (SAHAGIAN and STEER 1985; VON FIGURA et al. 1985) and was first purified by SAHAGIAN and coworkers (1981) using an affinity chromatography procedure. The complete primary sequence of the receptor has recently been deduced from cDNA clones encoding this protein (LOBEL et al. 1987; MORGAN et al. 1987; OSHIMA et al. 1988; LOBEL et al. 1988). The mature protein has a polypeptide backbone that is about 270 kDa in mass. After insertion into the ER membrane, the receptor requires as long as 2–3 h to properly fold and be exported to the Golgi complex (SAHAGIAN and NEUFELD 1983). This is significantly longer than the time required for a number of other membrane proteins to be exported from the ER (LODISH 1988; ROSE and DOMS 1988), and the difference may be due to the complexity of intrachain disulfide bond formation. In the Golgi complex, the receptor acquires complex-type oligosaccharide structures. Since approximately 20–30 kDa of the protein's mass is likely to be contributed by carbohydrate, the actual molecular weight of the cation-independent receptor is probably close to 300000.

A single, hydrophobic stretch divides the human 300-kDa receptor into exracellular and intracellular domains of 2265 and 164 amino acid residues respectively (MORGAN et al. 1987; OSHIMA et al. 1988). Analogous domains in the bovine receptor are 2269 and 163 residues in length (LOBEL et al. 1988). Nineteen potential asparagine-linked oligosaccharide addition sites are present in the extracellular domain of these proteins.

The entire extracellular portion of the 300-kDa receptor is comprised of 15 sequence repeat units, each about 145 residues in length, characterized by a conserved spacing of cysteine residues flanked by conserved hydrophobic residues. Each repeat unit contains a highly conserved 13-amino-acid stretch that matches the consensus sequence: Cys (Thr/Glu) Tyr X Phe (Glu) Trp X Thr X (Ala/Val) Ala Cys.[1] While the structural and functional significance of these repeats is not yet clear, it is important to note that the cysteine repeat unit is distinct from that seen in other receptors possessing cysteine-rich extracellular domains (YARDEN and ULLRICH 1988).

The 15 repeat units are interrupted only once, by a 43-amino-acid insertion which bears 53% sequence identity with the type II region of fibronectin. In fibronectin, this disulfide-bonded region contributes to a collagen-binding domain (KORNBLIHT et al. 1985). Tandem repeats of such sequences have also been detected in the blood clotting factor XII (MCMULLEN and FUJIKAWA 1985) and in a bovine seminal fluid protein (ESCH et al. 1983). Why the 300-kDa man6P receptor contains sequences homologous to fibronectin is not yet known.

The cytoplasmic domain is rather acidic and is the region most divergent among the bovine, rat, and human proteins. The cytoplasmic domain of this

[1] Residues shown in parentheses show a lower percentage match to the consensus; X denotes positions that are not conserved

receptor is almost twice the size of the corresponding domain of the cation-dependent man6P receptor (below) and that of other transport receptors such as those for low-density lipoproteins (LDL), transferrin, and asialoglycoproteins.

5 The Cation-Dependent Mannose 6-Phosphate Receptor

A number of cultured cell lines, including macrophages (P388D₁ and J774), fibroblasts (L cells), myeloma cells (MOPC315), and hepatoma cells (Morris hepatoma 7777), have been shown to lack the 300-kDa man6P receptor (GABEL et al. 1983; STEIN et al. 1987b). Despite the absence of the 300-kDa receptor, these cell lines are still able to direct a large portion of their lysosomal hydrolases to lysosomes. In their efforts to elucidate the basis for lysosomal enzyme targeting in such cells, HOFLACK and KORNFELD (1985a) identified the existence of a second, smaller man6P receptor, which appears to require divalent cations of binding to phosphomannosyl residue-containing lysosomal enzymes in vitro. These workers purified the cation-dependent man6P receptor to homogeneity from bovine liver and mouse P388D₁ cells; it is a glycosylated protein of about 46 kDa (HOFLACK and KORNFELD 1985b) and is probably a dimer (HOFLACK and KORNFELD 1985b; STEIN et al. 1987b). The deglycosylated protein has a molecular weight of approximately 28000; thus, a large portion of the receptor's mass is contributed by carbohydrate (HOFLACK and KORNFELD 1985b).

The primary sequences of both bovine and human 46-kDa receptors have recently been deduced from cDNA clones encoding these proteins (DAHMS et al. 1987; POHLMANN et al. 1987). The mature proteins are each comprised of 257 amino acids, and span the membrane once, exposing 67 amino acids to the cytoplasm. Five sites for potential asparagine-linked oligosaccharide addition are found clustered near the N-terminus, and biochemical studies indicate that at least four of these sites are modified (DAHMS et al. 1987; HOFLACK and KORNFELD 1985b; STEIN et al. 1987b). A very striking feature of the cation-dependent man6P receptor sequence is that the extracellular portion possesses a single copy of the same characteristic ~145-residue-long cysteine-rich structural repeat that appears 15 times in the 300-kDa receptor.

Direct comparison of the bovine and human cation-dependent receptor sequences reveals that they are 96% identical, differing at only 11 amino acid positions. All of these differences are clustered near the N terminus, which enables classification of receptor structural domains within the extracellular portion of the 46-kDa receptor. The first 28 amino acids of the mature proteins are identical. Then, a variable N-terminal region is followed by a cluster of closely spaced oligosaccharide addition sites. This glycosylated region is followed by 75 amino acids that are absolutely conserved between bovine and human receptors. As noted by DAHMS et al. (1987), it is this portion of the receptor that contains the most highly conserved, 13-amino-acid sequence that is present in each of the sequence repeats in the cation-independent man6P receptor. It is quite reasonable to propose that at least part of the ligand-binding domain will be localized to this membrane-proximal conserved region.

Considering the importance of the phosphate group in phosphomannosyl recognition by the 46-kDa receptor (HOFLACK et al. 1987), it would not be at all surprising if arginine and/or lysine groups were utilized for key ionic interactions with the phosphate moiety within the ligand-binding pocket. In the case of the enzyme phosphofructokinase, three arginine residues are used to form salt bridges with the phosphate group of fructose 6-phosphate. Indeed, STEIN and coworkers (1987a) have recently shown that arginine residues contribute to the ligand binding of the 46-kDa receptor. Modification of these residues with 1,2-cyclohexanedione abolished ligand binding; however, binding was much less severely affected if man6P was added prior to arginine modification to protect the ligand-binding site.

Comparison of the crystal structures of a number of carbohydrate-binding proteins has revealed the importance of hydrogen bonds in ligand binding (see QUIOCHO 1986 for review). The individual planar side chains of amino acids such as arginine, aspartic acid, glutamic acid, and asparagine are often utilized to form multiple hydrogen bonds with specific ligand sugars. The conserved membrane-proximal domain of the 46-kDa receptor is especially rich in this category of amino acids, which could be utilized both for the formation of salt bridges (arginine and lysine) and in hydrogen-bonding interactions. The availability of cDNA clones encoding both receptors will enable site-specific mutagenesis experiments to pinpoint the specific residues that are essential for ligand binding.

Unlike the cytoplasmic tail of the 300-kDa receptor, the 67-amino-acid cytoplasmic domains of the bovine and human 46-kDa receptors are absolutely conserved. They are comparable in size with the analogous domain of the LDL receptor (67 vs 50 residues). The cytoplasmic domain may function in homo-oligomeric interactions that might be required for accurate intracellular targeting between cellular compartments, as has been suggested for the LDL receptor (DAVIS et al. 1987; VAN DRIEL et al. 1987).

6 Intracellular Transport of Mannose 6-Phosphate Receptors

A number of laboratories have undertaken immunocytochemical and biochemical studies to determine the cellular distribution of the man6P receptors. First, it is clear that man6P receptors are not present in mature lysosomes by either biochemical (SAHAGIAN and NEUFELD 1983; VON FIGURA et al. 1984; STEIN et al. 1987b) or morphological (WILLINGHAM et al. 1981; GEUZE et al. 1984, 1985; BROWN et al. 1986) criteria. The receptors are found primarily in late endosomes; a small amount (less than 5%–10%) are located on cell surfaces.

Where do man6P receptors first encounter newly synthesized lysosomal enzymes? Man6P receptors present on the surface and within the cell are in rapid equilibrium, since over 90% contact extracellularly administered antibodies within a few hours (VON FIGURA et al. 1984; SAHAGIAN 1984; GARTUNG et al. 1985; PFEFFER 1987; NOLAN et al. 1987). DUNCAN and KORNFELD (1988) took advantage of this fact to monitor the transport of man6P receptors, labeled at the cell surface, back to the Golgi complex. These investigators found that both cation-dependent and cation-independent man6P receptors recycled back to the site of sialyltransferase (TGN and trans Golgi) with a half-time of about 3 h. Surprisingly, transport to the Golgi compartment that contains galactosyltransferase (the trans Golgi) was not detected. These experiments strongly suggest that man6P receptors bind to lysosomal hydolases in the TGN and sort them from secreted and plasma membrane proteins within this compartment (DUNCAN and KORNFELD 1988). These experiments also provide the first *functional* distinction between the trans Golgi and the TGN.

After binding lysosomal hydrolases in the TGN, man6P receptor-ligand complexes are collected into clathrin-coated pits which bud off from the TGN to form clathrin-coated vesicles (see VON FIGURA and HASILIK (1986) for review). These transport vesicles deliver the receptors to an acidic endosomal compartment, where lysosomal hydrolases are released (GONZALEZ-NORIEGA et al. 1980; SAHAGIAN 1984; BROWN et al. 1986). GRIFFITHS and coworkers (1988) and GEUZE et al. (1988) have recently completed detailed electron microscopic studies of the localization of the 300-kDa man6P receptor in normal rat kidney cells and rat hepatoma cells, respectively, using immunogold labeling procedures. These workers have indentified an "intermediate compartment" which both groups propose is the acidic delivery site for newly synthesized, receptor-bound lysosomal hydrolases en route to lysosomes. The intermediate compartment is located near the TGN, yet it is a distinct compartment, because it does not contain a viral glycoprotein that accumulates in the TGN at 20 °C (GRIFFITHS et al. 1988). Unlike the TGN, the intermediate compartment is highly enriched in both 300-kDa man6P receptors and lysosomal membrance glycoproteins (GRIFFITHS et al. 1988; GEUZE et al. 1988), and furthermore, the intermediate compartment appears to be significantly more acidic than the TGN (GRIFFITHS et al. 1988). At temperatures above 20 °C, the intermediate compartment receives endocytosed proteins. Taken together, these findings suggest that this newly identified compartment represents a prelysosome or "late endosome" structure.

Sorting of proteins takes place within the intermediate compartment, since man6P receptors are presumably retrieved from this compartment and returned to the Golgi complex. This is in contrast to our current notion of the events that occur within lysosomes. Because of this distinction, it may be most accurate to categorize the intermediate compartment as a type of late endosome.

Figure 2 summarizes our current model for the vesicular transport pathways taken by man6P receptors. These receptors shuttle primarily between the TGN and the so-called intermediate compartment (step a). Newly synthesized lysosomal enzymes bind to receptors in the TGN and are delivered to the

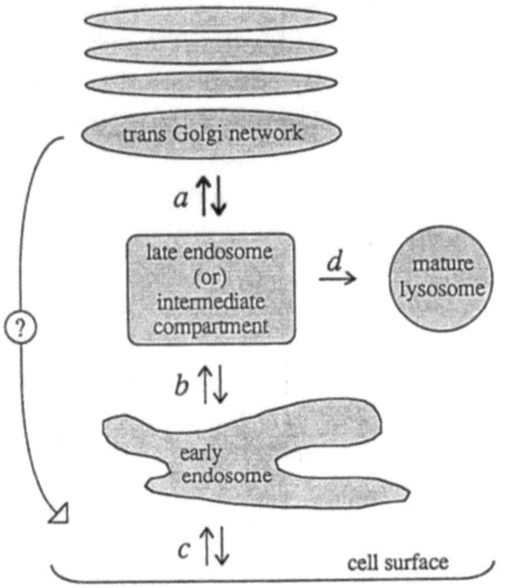

Fig. 2. Model for the intracellular transport of proteins en route to lysosomes. Man6P receptors bind to newly synthesized lysosomal enzymes in the trans-Golgi network (TEN) and carry them to an intermediate or late endosome compartment (step a). Ligands are released from receptors in this compartment due to the low intraorganellar pH, and man6P receptors recycle back to the TEN. A small number of man6P receptors are also present at the cell surface and can internalize extracellular lysosomal enzymes and deliver them to the intermediate compartment by passage through endosomes (steps *c, b*). Ligands delivered to the intermediate compartment, (*but not man6P receptors*), end up in mature lysosomes (step *d*). It is not yet known whether man6P receptors are ever transported directly from the TGN to the cell surface (*question mark*)

intemediate compartment; receptors then recycle back to the TGN for another round of transport. Man6P receptors can also bind to extracellular lysosomal enzymes at the cell surface and deliver them to the intermediate compartment (steps c, b). Lysosomal enzymes delivered to the intermediate compartment later appear in mature lysosomes (step d). This step may involve maturation of an intermediate compartment into a lysosome (HELENIUS et al. 1983).

Since all man6P receptors appear briefly at the cell surface, transport is likely to be possible from the intermediate compartment back to the cell surface (steps b, c). It is important to note, however, that transport from the intermediate compartment to early endosomes (step b) has not yet been demonstrated. The possibility that some map6P receptors are missorted and arrive at the cell surface directly from the TGN cannot yet be ruled out.

7 The Role of Mannose 6-Phosphate Receptors in Lysosomal Enzyme Targeting

Both the cation-dependent and cation-independent man6P receptors have been implicated directly in the recognition and targeting of man6P-containing proteins to lysosomes. As described above, man6P receptors are located primarily within the cell; however, they all appear briefly at the cell surface (and are then reinternalized) over a period of a few hours (VON FIGURA et al. 1984; SAHAGIAN 1984). When cells are incubated with antibodies that block the ligand binding sites of either receptor, the efficiency of lysosomal enzyme targeting is decreased

(VON FIGURA et al. 1984; GARTUNG et al. 1985; STEIN et al. 1987c; NOLAN et al. 1987). Under these conditions, a large proportion of man6P receptors come in contact with the antibodies, and lysosomal enzymes, unable to bind to man6P receptors, are secreted from the cells. These experiments strongly suggest that the 300-kDa and 46-kDa receptors both play a key role in lysosomal enzyme targeting.

Involvement of the 300-kDa receptor in lysosomal targeting has now been directly demonstrated in two laboratories. KYLE et al. (1988) expressed the human 300-kDa receptor in mouse $P388D_1$ cells which lack this protein, and showed that cation-independent receptor-expressing cells had a significantly higher ability to retain (and therefore sort) newly synthesized lysosomal hydrolases, as well as to endocytose exogenously added β-glucuronidase. LOBEL et al. (1989) obtained similar results using the bovine receptors expressed in a mouse L cell line which lacks the 300-kDa receptor.

In the case of the LDL receptor, naturally occurring mutations have greatly facilitated analyses of receptor function (GOLDSTEIN et al. 1985). No patients have yet been discovered with a defect in a man6P receptor (see KORNFELD 1986 for review); however, the introduction into cultured cells of in vitro-generated receptor mutants is already beginning to enrich our understanding of the cellular functions of the two man6P receptors.

8 The Mannose 6-Phosphate Receptor Cytoplasmic Domain Is Important for Sorting

LOBEL et al. (1989) have used recombinant DNA techniques to create a series of 300-kDa man6P receptor mutants, and they have examined the functions of the mutant proteins after expression in receptor-negative, cultured cells. Deletion of almost 90% of the 163-amino-acid cytoplasmic domain generated a receptor that could bind lysosomal enzymes, but was poorly endocytosed, accumulated at the cell surface, and failed to sort lysosomal hydrolases. In contrast, deletion of only about one-quarter or one-half of the cytoplasmic domain generated receptors that were capable of "wild type" levels of endocytosis but displayed defects in lysosomal enzyme sorting. These experiments demonstrate that the membrane-proximal portion of the cytoplasmic domain is essential for endocytosis. Furthermore, LOBEL et al. have shown that the C-terminal-most portion of the man6P receptor cytoplasmic domain possesses structural information that is important for receptor trafficking. It is not unreasonable to postulate that cytosolic factors recognize features of the cytoplasmic domain and in some way expedite the collection of man6P receptors into clathrin-coated transport vesicles.

Pearse (1988) has proposed that so-called "adaptor" proteins, constituents of the clathrin coat, bind directly to receptor cytoplasmic domains. Using purified adaptors and affinity matrices of receptor cytoplasmic domains, PEARSE and coworkers have reported detection of precisely this type of binding (PEARSE 1988; GLICKMAN et al. 1989). These workers concluded that purified adaptors derived from either plasma membrane or Golgi coated pits bound to independent sites on the 300-kDa man6P receptor cytoplasmic domain. One difficulty with these experiments was the fact that significant binding (up to 50%) was observed using control columns. However, in conjunction with the appropriate mutant cytoplasmic domains, this powerful approach may yet reveal the cytosolic factors that facilitate collection of man6P receptors into clathrin-coated transport vesicles.

DINTZIS and PFEFFER (1990) have obtained evidence suggesting that the cytoplasmic domain is not sufficient to direct the routing of the 300-kDa man6P receptor. These workers constructed a chimeric receptor in which the extracellular and transmembrane domains of human EGF receptor were linked precisely to the cytoplasmic domain of the bovine 300-kDa man6P receptor; they then expressed this protein in cultured cells to examine its transport. If the cytoplasmic domain contains all the information needed to bind adaptors (or other cytosolic factors) and determine receptor targeting, the chimera would be expected to share the same cellular distribution as the endogenous man6P receptor: primarily within late endosomes. Surprisingly, greater than 90% of the chimera was localized to the cell surface, under conditions in which greater than 90% of the endogenous 300-kDa man6P receptor was intracellular. The chimera was not stuck at the surface, because it bound EGF with high affinity, it internalized the ligand, and it recycled continuously for over 2h. Thus, the man6P receptor cytoplasmic domain was not sufficient to alter the steady state distribution of the EGF receptor. The predominance of the chimera on the cell surface makes it difficult to imagine that this protein shuttles between late endosomes and the TGN at any appreciable rate.

From these experiments, one can infer that extracellular (or luminal) portions of the 300-kDa man6P receptor are responsible for its primarily intracellular localization. Consistent with this conclusion, 80% of the endocytosis-competent mutant receptors lacking a portion of the cytoplasmic domain were still located intracellularly, despite a defect in lysosomal enzyme sorting (LOBEL et al. 1989). In summary, it appears that the cytoplasmic domain in necessary, but not sufficient, for lysosomal enzyme targeting.

9 Comparison of the Two Mannose 6-Phosphate Receptors

It is not yet clear why cells have two types of man6P receptors. Once could have imagined that the two proteins carried out different functions, bound different classes of lysosomal enzymes, or were present in different cell types or cellular locations. At present, none of these possibilities appears to be correct.[2] The 46-kDa and 300-kDa receptors share a broad tissue distribution, and while the two receptors are present in roughly equimolar amounts in bovine liver (HOFLACK and KORNFELD 1985b), the 46-kDa receptor is the predominant phosphomannosyl-binding protein in homogenates of bovine tests (DISTLER and JOURDIAN 1987; DISTLER et al. 1987). A few cell lines lack the 300-kDa man6P receptor (see above), but no cell line has yet been identified that lacks the 46-kDa receptor, and most cell types appear to contain both.

As described above, the 46-kDA and 300-kDa man6P receptors are both specific for oligosaccharides bearing phosphomannosyl groups in monoester linkage, and bind with highest affinity to oligosaccharides containing two such moieties. Both proteins release their ligands at pH values below 5.5, a characteristic feature of receptors that deliver their cargo to acidic intracellular compartments (SAHAGIAN et al. 1981; HOFLACK and KORNFELD 1985a, b). HOFLACK et al. (1987) were unable to detect any differences between the two receptors with regard to their ability to bind to any particular subset of lysosomal enzymes. However, the 46-kDa receptor can be distinguished from the 300-kDa receptor in that it does not bind to methylphosphomannosyl groups present on *Dictyostelium discoideum* lysosomal enzymes (HOFLACK and KORNFELD 1985a), and it displays a distinct pH profile for binding to lysosomal enzymes in vitro (HOFLACK et al. 1987).

Despite similar binding properties, the primary sequences of the 46-kDa and 300-kDa receptors do not display a high degree of sequence identity; the only apparent common feature is the presence of the ~145-amino-acid cysteine-rich repeat units. Yet, the repeat units appear not to represent individual man6P-binding domains, since the 300-kDa receptor and a 46-kDa receptor dimer each bind 2 mol of man6P (KORNFELD 1987), far less than the number of repeat units (15) present in the 300-kDa receptor. Variable numbers of another type of cysteine-rich repeat occur in other cell surface receptors, including those for LDL, EGF, and insulin (PFEFFER and ULLRICH 1986). Since multiple copies of this cysteine-rich repeat are found in cell surface receptors with very diverse functions, they appear to form an important structural element in the vicinity of ligand-binding domains, *without* defining binding specificity. It is still unclear why two proteins with the same apparent function have such disparate structures.

---

[2] The precise intracellular distribution of the 46-kDa receptor, as determined by electron microscopy, has not yet been reported. However, biochemical experiments by STEIN et al. (1987b) and by DUNCAN and KORNFELD (1988) indicate that its distribution is likely to be very similar to that of the 300-kDa receptor

10 The 300-kDa Mannose 6-Phosphate Receptor is Identical to the Insulin-Like Growth Factor Type II Receptor

The deduced amino acid sequence of the insulin-like growth factor type II (IGF-II) receptor has recently been obtained from full-length cDNA clones (MORGAN et al. 1987). Quite unexpectedly, the human sequence was found to be over 80% identical with the sequence of the bovine cation-independent man6P receptor (LOBEL et al. 1987, 1988) and, more recently, over 99% identical with the sequence of the human cation-independent man6P receptor (OSHIMA et al. 1988). Does one of these cloned receptors have a mistaken identity? Apparently not. First, in each case, the isolated cDNA clones contained extensive, known peptide sequences derived from independently purified IGF-II and 300-kDa man6P receptor preparations. MORGAN and coworkers (1987) expressed their cloned IGF-II receptor cDNA in oocytes and observed the appearance of IGF-II binding sites on the surfaces of these cells. Expression of the human man6P receptor cDNA generated a protein that could mediate β-glucuronidase endocytosis (OSHIMA et al. 1988). Thus, these essentially identical cDNAs encode a protein capable of binding both IGF-II and lysosomal enzymes.

A number of laboratories have now shown that a single polypeptide chain has the ability to bind both IGF-II and man6P, and that the binding sites for these ligands are distinct (ROTH et al. 1987; TONG et al. 1988; MACDONALD et al. 1988; WAHEED et al. 1988; KIESS et al. 1988). The 300-kDa man6P/IGF-II receptor would be expected to contain two distinct binding sites, since the binding of lysosomal enzymes to this receptor absolutely requires the presence of man6P groups, and IGF-II is not even glycosylated. In hindsight, a review of the characteristics of the protein studied either as the "IGF-II receptor" or as the "cation-independent man6P receptor" reveals numerous similarities, including similar apparent molecular weights, high disulfide bridge content, broad tissue distribution, and primarily intracellular localization.

Two groups have reported that the affinity of the 300-kDa receptor for IGF-II increases twofold in the presence of man6P (MACDONALD et al. 1988; ROTH et al. 1987). These experiments were carried out using a receptor that was purified on an IGF-II affinity column. Using receptor purified on a phosphomannan affinity column, TONG et al. (1988) detected no effect of one ligand on the binding of the other. If any contaminating lysosomal enzymes interfered sterically with subsequent IGF-II binding, addition of man6P would release the enzyme and yield an apparent increase in IGF-II binding. Indeed, KIESS et al. (1988) have shown that IGF-II binding is inhibited by the presence of β-galactosidase, and can be reversed by addition of man6P. Collectively, these data point to a lack of cooperativity between the two binding sites.

The presence of distinct binding sites of IGF-II and man6P-containing proteins may in part explain why the 300-kDa receptor is larger than the 46-kDA man6P receptor, which does not bind IGF-II (TONG et al. 1988). The possibility remains that the 300-kDa receptor binds yet another ligand (or ligands), in addition to IGF-II and proteins that contain man6P.

11 Significance of a Multifunctional Receptor

In bacterial chemotaxis, a single receptor senses the concentrations of both aspartate and maltose (bound to the maltose-binding protein) and can trigger a chemotactic response to either or both of these components independently and additively (MOWBRAY and KOSHLAND 1987). In this case, bacterial cells use a single receptor to couple two ligands to the same chemotactic signalling machinery. Might there also be physiological consequences of binding both IGF-II and lysosomal enzymes to a single receptor? Alternatively, do the two binding sites reflect independent functions of this receptor protein? Although the 300-kDa receptor appears to play a key role in targeting enzymes to lysosomes (described above), the significance of this molecule as an IGF-II-binding protein is not yet fully established (see ROTH 1988 for review).

The first issue that must be considered is whether the man6P/IGF-II receptor transduces a growth factor signal. If the receptor transmits a signal across the plasma membrane, it would be all the more difficult to reconcile the cell's use of the same receptor protein to target proteins to lysosomes and to generate an intracellular signal. The importance of this issue becomes clear when one considers the complexities of ligand interaction with insulin and IGF receptors.

Insulin and the related growth factors, IGF-I and IGF-II, are structurally homologous growth factors that bind to specific, cognate cell surface receptors (see CZECH 1982; RECHLER and NISSLEY 1985 for review). These growth factors also bind to heterologous members of this receptor family with somewhat lower affinites in vitro. Thus, IGF-I can interact with both insulin and IGF-II receptors, and insulin and IGF-II can also bind to the IGF-I receptor.[3] Insulin and IGF-I receptors are highly homologous proteins that possess an intrinsic tyrosine kinase activity that is essential for signal transduction (ULLRICH et al. 1985; EBINA et al. 1985; ULLRICH et al. 1986). In contrast, the primary sequence of the IGF-II receptor is completely unrelated to that of the insulin and IGF-I receptors (MORGAN et al. 1987), and the IGF-II receptor lacks tyrosine kinase activity in vitro. The absence of sequence homology between IGF-I and IGF-II receptors suggests that the IGF-II binding sites present on both of these proteins either interact with different portions of the IGF-II molecule or alternatively consist of similarly folded protein domains.

Since IGF-II can activate the tyrosine kinase activity of the IGF-I receptor, it has been extremely difficult to dissociate the physiological effects of IGF-II that are mediated by its own receptor from those triggered by the IGF-I receptor. Thus, while several reports have suggested that the IGF-II receptor does indeed transduce a signal (NISHIMOTO et al. 1987a, b; HARI et al. 1987; TALLY et al. 1987), other investigators have concluded that it does not (MOTTOLA and CZECH 1984; KRETT et al. 1987).

[3] IGF-II can apparently also interact with the insulin receptor, but only a very high concentrations in vitro

The simplest explanation for the presence of two ligand-binding sites on the 300-kDa receptor is that this receptor is merely used to clear IGF-II from the circulation. For unknown reasons, IGF-II is present at rather high levels (300-600 ng/ml, or ~60 n*M*) in human serum. By internalizing IGF-II and releasing it within endosomes, the 300-kDa receptor would target IGF-II to lysosomes for degradation. In this model, a different protein would be used to mediate an IGF-II signal: either the IGF-I receptor or another, yet to be identified IGF-II binding protein.

If the man6P/IGF-II receptor does indeed transduce a growth factor signal, the consequences of intracellular stimulation of the receptor, in cell types such as liver that express both the receptor and its cognate growth factor, must be considered. However, the cytoplasmic tail of a signal-transducing receptor might be expected to have been more highly conserved.

12 Mannose 6-Phosphate-Independent Sorting to Lysosomes

The analysis of man6P receptors has uncovered a very satisfying pathway for lysosomal enzyme sorting, yet it is clear that man6P receptors accomplish only part of this sorting task. An indication of the existence of a man6P-independent sorting pathway has come from the analysis of cells from patients with I-cell disease. As described earlier, cells from such patients are unable to construct man6P-containing oligosaccharides because they lack phosphotransferase activity. Paradoxically, while I-cell fibroblasts are deficient in several lysosomal enzymes (see above), intracellular levels of lysosomal enzymes are near normal in liver, spleen, kidney, and brain (WAHEED et al. 1982; OWADA and NEUFELD 1982). The basis for this targeting is presently unknown.

Recent characterization of a number of membrane-associated lysosomal proteins, including a family of lysosomal membrane glycoproteins (lgp) of unknown function (LEWIS et al. 1985; CHEN et al. 1985; BARRIOCANAL et al. 1986; LIPPINCOTT-SCHWARTZ and FAMBROUGH 1986), has revealed that none of these proteins acquire man6P groups on their oligosaccharide side chains. Therefore, some other mechanism must account for their accurate targeting to lysosomes. It is tempting to speculate that the same sorting machinery responsible for the targeting of man6P receptors (which themselves lack man6P residues) to the intermediate compartment may also segregate lgps and target them to this destination (GREEN et al. 1987). Whatever signals are involved in this sorting must be unique to this class of membrane proteins.

The nucleotide sequences of several members of this family of lgp are now available, and the sequences divulge an interesting clue as to the signal that may direct these proteins to lysosomes. Analysis of lgp sequences from chicken (FAMBROUGH et al. 1988), mouse (CHEN et al. 1988), human (VIITALA et al. 1988;

FAKUDA et al. 1988), and rat cells (HOWE et al. 1988) has revealed that each of these related gene products bears an identical, 11-amino-acid cytoplasmic domain. The cytoplasmic and transmembrane domains are the most highly conserved portions of lgps, and are entirely unrelated to the corresponding domains of either of the man6P receptors. Initial analysis of the function of the lgp cytoplasmic domain already indicates that it plays an important role in the targeting of lgps to lysosomes (MATHEWS and FAMBROUGH 1988).

13 Studies of Mannose 6-Phosphate Receptor Transport in a Cell-Free System

Since little is known about how man6P receptors are collected into transport vesicles and how such transport vesicles form, identify their targets, and then fuse, GODA and PFEFFER (1988) sought to reconstitute the vesicular transport of man6P receptors between cellular compartments with the goal of elucidating these transport reactions in molecular terms. To do this, they took advantage of the semi-intact cell preparation described by BECKERS et al. (1987) in which swollen cells are scraped from a monolayer. This process removes portions of the plasma membrane, and gently perforates them to enable analysis of vesicular transport (see also SIMONS and VIRTA 1987).

GODA and PFEFFER (1988) devised a complementation scheme to monitor the transport of the 300-kDa man6P receptor from late endosomes to the TGN. The scheme takes advantage of the fact that sialyltransferase is located exclusively in the trans Golgi and TGN (ROTH et al. 1985). Thus, the addition of sialic acid to man6P receptor oligosaccharides can be used to mark the arrival of this receptor in these Golgi compartments. To facilitate detection, the transport assay also utilizes a mutant cell line, CHO^{1021}, which displays an apparent defect in sialyltransferase activity. In CHO^{1021} cells, man6P receptor asparangine-linked oligosaccharides terminate in galactose and are excellent substrates for sialyltransferase.

The assay is carried out as follows. CHO^{1021} cells are labeled with ^{35}S-methionine and then gently broken, and aliquots of the cell extract are mixed with "wild-type" rat liver Golgi complexes, a crude cytosol fraction, ATP, and an ATP-regenerating system. The mixtures are incubated at 37 °C, after which man6P receptors are isolated by affinity chromatography. If CHO^{1021} cell man6P receptors are transported to the wild-type trans Golgi or TGN, radiolabled receptors should acquire sialic acid, which is monitored by chromatography of the isolated receptors on a column of the sialic acid-specific *Limax flavus* slug lectin. Transported receptors bind to such columns and elute in the presence of excess sialic acid; nontransported proteins bearing galactose-terminating oligosaccharides do not bind, and are collected in the flow-through fractions. Samples are then analyzed by direct scintillation counting and SDS-PAGE.

As would be expected for a vesicular transfer, transport was entirely time-, temperature-, and ATP-dependent, the components used to monitor transport resided in sealed membrane compartments throughout the reaction, and control experiments strongly suggested that the transported receptor originated in a late endosome compartment. The physiological significance of the process measured in vitro was supported both by its efficiency and by its apparent selectivity: as observed in living cells (JIN et al. 1989), transferrin receptors recycled to the TGN at a lower rate than man6P receptors in the cell-free system. Furthermore, no receptor transport was observed at 18 °C, as is true in vivo (SNIDER and ROGERS 1985; DUNCAN and KORNFELD 1988). Recent experiments demonstrate that transport depends upon the addition of crude cytosol (GODA and PFEFFER 1991), analogous to other vesicular transport events (PFEFFER and ROTHMAN 1987).

Similar to intra-Golgi transport (MELANCON et al. 1987), recycling of the man6P receptor to the TGN required the hydrolysis of GTP. It is probable that GTP plays a regulatory role in these processes, and it will be interesting to determine whether these distinct transport steps are regulated by common GTP binding proteins. In yeast, proteins related to *ras* have been implicated both in Golgi transport (SEGEV et al. 1988) and in secretory vesicle exocytosis (SALMINEN and NOVICK 1987). The increasing number of identified G proteins and *ras*-related proteins raises the alternative possibility that individual members of a family of such proteins facilitate distinct transport events.

Clathrin-coated vesicles retrieve man6P receptors from the cell surface, and also carry them out of the TGN. Yet the selectivity of these classes of transport vesicles must be quite distinct, since the same nutrient and signalling receptors that are collected together with man6P receptors into coated pits at the cell surface (WILLINGHAM et al. 1981) must be sorted away from man6P receptors in the TGN (GRIFFITHS and SIMONS 1986). Little is known about the nature of the vesicles that carry man6P receptors from late endosomes to the TGN, and the recent discovery of coated transport vesicles that do not contain clathrin (GRIFFITHS et al. 1985; ORCI et al. 1986) opens the possibility that clathrin may not be involved. This is supported by the observation that anti-clathrin antibodies fail to block endosome-TGN transport, yet inhibit endocytosis in vitro (DRAPER et al. 1990).

14 Perspectives

The existence of man6P-independent pathways for lysosomal enzyme targeting emphasizes the fact that man6P receptors are themselves not responsible for the actual lysosomal enzyme sorting process. Instead, these receptors represent an efficiency mechanism by which the cell can couple perhaps as many as 50 different lysosomal enzymes to a single, yet to be identified sorting machinery. A major challenge for the future will be to identify the molecular components of this sorting machinery and to establish how the cell segregates man6P receptors and lysosomal membrane glycoproteins from proteins bound for other cellular destinations. The availability of cDNA clones encoding man6P receptors and lysosomal membrane glycoproteins, as well as cell-free systems which reconstitute man6P receptor transport, will undoubtedly accelerate our progress towards elucidating these fundamental processes.

References

Barriocanal JG, Bonifacino JS, YUAN L, Sandoval IV (1986) Biosynthesis, glycosylation, movement through the Golgi system, and transport to lysosomes by an N-linked carbohydrate-independent mechanism of three lysosomal intergral membrane proteins. J Biol Chem 261: 16755–16763

Beckers CJM, Keller DS, Balch WE (1987) Semi-intact cells permeable to macromolecules: use in reconstitution of protein transport from the endoplasmic reticulum to the Golgi complex. Cell 50: 523–534

Brown WJ, Goodhouse J, Farquhar MG (1986) Mannose 6-phosphate receptors for lysosomal enzymes cycle between the Golgi complex and endosomes. J Cell Biol 103: 1233–1247

Chen JW, Murphy TL, Willingham MC, Pastan I, August JT (1985) Identification of two lysosomal membrane glycoproteins. J Cell Biol 101: 85–95

Chen JW, Cha Y, Yuksel KU, Gracy RW, August JT (1988) Isolation and sequencing of a cDNA clone encoding lysosomal membrane glycoprotein mouse LAMP-1: sequence similarity to proteins bearing onco-differentiation antigens. J Biol Chem 263: 8754–8758

Creek KE, Sly WS (1982) Adsorptive pinocytosis of phosphorylated oligosaccharides by human fibroblasts. J Biol Chem 257: 9931–9937

Czech MP (1982) Structural and functional homologies in the receptors for insulin and the insulin-like growth factors. Cell 31: 8–10

Dahms NM, Lobel P, Breitmeyer J, Chirgwin JM, Kornfeld S (1987) 46kD mannose 6-phosphate receptor:cloning, expression, and homology to the 215 kD mannose 6-phosphate receptor. Cell 50: 181–192

Davis CG, van Driel IR, Russel DW, Brown MS, Goldstein JL (1987) The low density lipoprotein receptor: identification of amino acids in the cytoplasmic domain required for rapid endocytosis. J Biol Chem 262: 4075–4082

Deutscher SL, Creek KE, Merion M, Hirschberg CB (1983) Subfractionation of rat liver Golgi apparatus: separation of enzyme activities involved in the biosynthesis of the phosphomannosyl recognition marker in lysosomal enzymes. Proc Natl Acad Sci USA 80: 3938–3942

Dintzis SM, Pfeffer SR (1990) The mannose 6- phosphate receptor cytoplasmic domain is not sufficient to alter the cellular distribution of a chimeric EGF receptor EMBO Jl 9: 77–84

Distler JJ, Jourdian GW (1987) Low molecular weight phosphomannosyl receptor from bovine testes. Methods Enzymol 138: 504–509

Distler JJ, Patel R, Jourdian GW (1987) Immobilization and assay of low-molecular-weight phosphomannosyl receptor in multiwell plates. Anal Biochem 166: 65–71

Draper RK, Goda Y, Brodsky FM, Pfeffer SR (1990) Antibodies to clathrin inhibit endocytosis but not recycling to the trans Golgi network in vitro. Science 248: 1539–1541

Duncan J, Kornfeld S (1988) Intracellular movement of two mannose 6-phosphate receptors:return to the Golgi aparatus. J Cell Biol 106: 617–628

Ebina Y, Ellis L, Jarnagin K, Edery M, Graf L, Clauser E, Ou J-H, Masiarz F, Kan YW, Goldfine ID, Roth RA, Rutter WJ (1985) The human insulin receptor cDNA: the strucutral basis for hormone-activated transmembrane signalling. Cell 40: 747–758

Esch FS, Ling NC, Bohlen P, Ying SY, Guillemin R (1983) Primary structure of PDC-109, a major protein constituent of bovine seminal plasma. Biochem Biophys Res Commun 113: 861–867

Fambrough DM, Takeyasu K, Lippincott-Schwartz J, Siegel NR (1988) Structure of LEP100, a glycoprotein that shuttles between lysosomes and the plasma membrane, deduced from the nucleotide sequence of the encoding cDNA. J Cell Biol 106: 61–67

Faust PL, Chirgwin JM, Kornfeld S (1987) Renin, a secretory glycoprotein, acquires phosphomannosyl residues. J Cell Biol 105: 1947–1955
Fischer HD, Creek KE, Sly WS (1982) Binding of phosphorylated oligosaccharides to immobilized phosphomannosyl receptors. J Biol Chem 257: 9938–9943
Fukuda M, Viitala J, Matteson J, Carlsson SR (1988) Cloning of CDNAs encoding human lysosomal membrane glycoprotiens, H-lamp-1 and h-lamp-2: comparison of their deduced sequences. J Biol Chem 263: 18920–18928
Gabel, CA, Goldberg DE, Kornfeld S (1983) Identification and characterization of cells deficient in the mannose 6-phosphate receptor: evidence for an alternate pathway for lysosomal enzyme targeting. Proc Natl Acad Sci USA 80: 775–779
Gartung C, Braulke T, Hasilik A, von Figura K (1985) Internationalization of blocking antibodies against mannose 6-phosphate specific receptors. EMBO J 4: 1725–1730
Geuze HJ, Slot JW, Strous GJAM, Hasilik A, von Figura K (1984) Ultrastructural localization of the mannose 6-phosphate receptor in rat liver. J Cell Biol 98: 2047–2054
Geuze JH, Slot JW, Strous GJAM, Hasilik A, von Figura K (1985) Possible pathways for lysosomal enzyme delivery. J Cell Biol 101: 2253–2262
Geuze HJ, Stoorvogel W, Strous GJ, Slot JW, Zijderhand-Bleekemolen J, Mellman I (1988) The sorting of mannose 6-phosphate receptors and lysosomal membrane proteins occurs in endocytic vesicles. J Cell Biol 107: 2491–2502
Glickman JN, Conibear E, Pearse BMF (1989) Specificity of binding clathrin adaptors to signals on the mannose 6-phosphate/insulin-like growth factor II receptor. EMBO J 8: 1041–1047
Goda Y, Pfeffer SR (1988) Selective recycling of the mannose 6-phosphate/IGF-II receptor to the trans Golgi network in vitro. Cell 55: 309–320
Goda Y, Pfeffer SR (1991) Identification of a novel, N-ethylmaleimide-sensitive cytosolic factor required for vesicular transport from endosomes to the trans Golgi network in vitro. J Cell Biol 112: 823–831
Goldberg DE, Kornfeld S (1981) The phosphorylation of β-glucuronidase oligosaccharides in mouse P388 D_1 cells. J Biol Chem 256: 13060–13067
Goldberg DE, Kornfeld S (1983) Evidence for extensive subcellular organization of asparagine-linked oligosaccharide processing and lysosomal enzyme phosphorylation. J Biol Chem 258: 3159–3165
Goldstein JL, Brown MS, Anderson RGW, Russell DW, Schneider WJ (1985) Receptor-mediated endocytosis: concepts emerging from the LDL receptor system. Annu Rev. Cell Biol 1: 1–39
Gonzalez-Noriega A, Grubb JH, Talkad V, Sly WS (1980) Chloroquine inhibits lysosomal enzyme pinocytosisand enhances lysosomal enzyme secretion by impairing receptor recycling. J Cell Biol 85: 839–852
Green SA, Zimmer K-P, Griffiths G, Mellman I (1987) Kinetics of intracellular transport and sorting of lysosomal membrane and plasma membrane proteins. J Cell Biol 105: 1227–1240
Griffiths G, Simons K (1986) The trans Golgi network: sorting at the exit site of the Golgi complex. Science 234: 438–443
Griffiths G, Pfeiffer S, Simons K, Matlin K (1985) Exit of newly synthesized membrane proteins from the trans cisternae of the Golgi complex to the plasma membrane. J Cell Biol 101: 949–964
Griffiths G. Hoflack B, Simons K. Mellman I, Kornfeld S (1988) The mannose 6-phosphate receptor and the biogenesis of lysosomes. Cell 52: 329–341
Hari J, Pierce SB, Morgan DO, Sara V, Smith MC, Roth RA (1987) The receptor for insulin-like growth factor II mediates an insulin-like response. EMBO J 6: 3367–3371
Helenius A, Mellman I, Wall D, Hubbard A (1983) Endosomes. Trends Biochem Sci 8: 245–250
Hoflack B, Kornfeld S (1985a) Lysosomal enzyme binding to mouse P388D1 macrophage membranes lacking the 215kDa mannose 6-phosphate receptor: evidence for the existence of a second mannose 6-phosphate receptor. Proc Natl Acad Sci USA 82:4428–4432
Hoflack B, Kornfeld S (1985b) Purification and characterization of a cation-dependent mannose 6-phosphate receptor form murine P388D1 macrophages and bovine liver. J Biol Chem 260: 12008–12014
Hoflack B, Fujimoto K, Kornfeld S (1987) The interaction of phosphorylated oligosaccharides and lysosomal enzymes with bovine liver cation-dependent mannose 6-phosphate receptor. J Biol Chem 262: 123–129
Howe CL,Granger BL, Hull M, Green SA, Gabel CA, Helenius A, Mellman I (1988) Derived protein sequence, oligosaccharides, and membrane insertion of the 120-kDa lysosomal membrane glycoprotein (lgp 120): identification of a highly conserved family of lysosomal membrane glycoproteins. Proc Natl Acad Sci USA 85: 7577–7581

Jin MJ, Sahagian GG, Sinder MD (1989) Transport of surface mannose 6-phosphate receptor to a sialylatransferase-containing compartment in cultured human cells. J Cell Biol 108 (in press)
Kaplan A, Achord DT, Sly WS (1977) Phosphohexosyl components of a lysosomal enzyme are recognized by pinocytosis receptors on human fibroblasts. Proc Natl Acad Sci USA 74: 2026–2030
Kiess W, Blickenstaff GD, Sklar MM, Thomas CL, Nissley SP, Sahagian GG (1988) Biochemical evidence that the type II insulin-like growth factor receptor is identical to the cation-independent mannose 6-phosphate receptor. J Biol Chem 263: 9339–9344
Kornblihit AR, Umezawa K, Vibe-Petersen K, Baralle F (1985) Primary structure of human fibronectin: differential splicing may generate at least ten polypeptides from a single gene. EMBO J 4: 1755–1759
Kornfeld R, Kornfeld S (1985) Assembly of asparagine-linked oligosaccharides. Annu Rev Biochem 54: 631–664
Kornfeld S (1986) Trafficking of lysosomal enzymes in normal and disease states. J Clin Invest 77: 1–6
Kornfeld S (1987) Trafficking of lysosomal enzymes. FASEB J 1: 462–468
Kornfeld S, Mellman I (1989) The biogenesis of lysosomes. Annu Rev Cell Biol 5: 483–525
Krett NL, Heaton JH, Gelehrter TD (1987) Mediation of insulin-like growth factor actions by the insulin receptor in H-35 rat hepatoma cells. Endocrinology 120: 401–408
Kyle JW, Nolan CM, Oshima A, Sly WS (1988) Expression of human cation-independent mannose 6-phosphate receptor cDNA in receptor-negative mouse $P388D_1$ cells following gene transfer. J Biol Chem 263: 16230–16235
Lang L, Reitman ML, Tang J, Roberts RM, Kornfeld S (1984) Lysosomal enzyme phosphorylation. Recognition of a protein-dependent determinant allows specific phosphorylation on oligosaccharides present on lysosomal enzymes. J Biol Chem 259: 14663–14671
Lang L, Takahashi T, Tang J, Kornfeld S (1985) Lysosomal enzyme phosphorylation in human fibroblasts: kinetic parameters offer a biochemical rationale for two distinct defects in the UDP-GlcNAc: lysosomal enzyme precursor N-acetylglucosamine 1-phosphotransferase. J Clin Invest 76: 2191–2195
Lazzarino DA, Gabel CA (1988) Biosynthesis of the mannose 6-phosphate recognition marker in transport-impaired mouse lymphoma cells: demonstration of a two-step phosphorylation. J Biol Chem 263: 10118–10126
Kee S-J, Nathans D (1988) Proliferin secreted by cultured cells binds to mannose 6-phosphate receptors. J Biol Chem 263: 3521–3527
Lewis V, Green SA, Marsh M Vikho P, Helenius A, Mellman I (1985) Glycoproteins of the lysosomal membrane. J Cell Biol 100: 1839–1847
Lippincott-Schwartz J, Fambrough DM (1986) Lysosomal membrane dynamics: structure and interorganellar movement of a major lysosomal membrane glycoprotein. J Cell Biol 102: 1593–1605
Lobel P, Dahms NM, Breitmeyer J, Chirgwin JM, Kornfeld S (1987) Cloning of the bovine 215-kDa cation-independent mannose 6-phosphate receptor. Proc Natl Acad Sci USA 84: 2233-2237
Lobel P, Dahms NM, Kornfeld S (1988) Cloning and sequence analysis of the cation-independent mannose 6-phosphate receptor. J Biol Chem 263: 2563–2570
Lobel P, Fujimoto K, Ye RD, Griffiths G, Kornfeld S (1989) Mutations in the cytoplasmic domain of the 275 kd mannose 6-phosphate receptor differentially alter lysosomal enzyme sorting and endocytosis. Cell (in press)
Lodish H (1988) Transport of secretory and membrane glycoproteins from the rough endoplasmic reticulum to the Golgi: a rate limiting step in protein maturation and secretion. J Biol Chem 263: 2107–2110
MacDonald RG, Pfeffer SR, Coussens L, Tepper MA, Brocklebank CM, Mole JE, Anderson JK, Chen E, Czech MP, Ullrich A (1988) A single receptor binds both IGF-II and mannose 6-phosphate. Science 239: 1134–1137
Mathews PM, Fambrough DM (1988) Targeting sequences in the shuttling membrane glycoprotein LEP100 sufficient for lysosomal distribution. J Cell Biol 107: 439a
Mc Mullen BA, Fujikawa K (1985) Amino acid sequence of the heavy chain of human α-factor XIIa (activated Hageman factor). J Biol Chem 260: 5328–5341
Melancçon P, Glick BS, Malhotra V, Weidman PJ, Serafini T, Gleason ML, Orci L, Rothman JE (1987) Involvement of GTP-binding "G" proteins in transport through the Golgi stack. Cell 51: 1053–1062
Morgan DO, Edman JC, Standring DN, Fried VA, Smith MC, Roth RA, Rutter WJ (1987) Insulin-like growth factor II receptor as a multifunctional binding protein. Nature 329: 301–307

Mottola C, Czech MP (1984) The type II insulin-like growth factor receptor does not mediate increased DNA synthesis in H-35 hepatoma cells. J Biol Chem 259: 12705–12713
Mowbray SL, Koshland DE (1987) Additive and independent responses in a single receptor: aspartate and maltose stimuli on the Tar protein. Cell 50: 171–180
Nishimoto I, Ohkuni E, Ogata E, Kojima I (1987a) Insulin-like growth factor II increases cytoplasmic free calcium in competent Balb/c 3T3 cells treated with epidermal growth factor. Biochem Biophys Res Commun 142: 275–282
Nishimoto I, Hata Y, Ogata E, Kojima I (1987b) Insulin-like growth factor II stimulates calcium influx in competent Balb/c 3T3 cells primed with epidermal growth factor: characteristics of calcium influx and involvement of GTP-binding protein. J Biol Chem 262: 12120–12126
Nolan CM, Creek KE, Grubb J, Sly WS (1987) Antibody to the phosphomannosyl receptor inhibits recycling of receptor in fibroblasts. J Cell Biochem 35: 137–151
Orci L, Glick BS, Rothman JE (1986) A new type of coated vesicular carrier that appears not to contain clathrin: its possible role in protein transport within the Golgi stack. Cell 46: 171–184
Oshima A, Nolan C, Kyle JW, Grubb JH, Sly WS (1988) The human cation-independent mannose 6-phosphate receptor: cloning and sequence of full-length cDNA and expression of functional receptor in COS cells. J Biol Chem 263: 2553–2562
Owada M, Neufeld EF (1982) Is there a mechanism for introducing acid hydrolases into liver lysosomes that is independent of mannose 6-phosphate recognition? Biochem Biophys Res Commun 105: 814–820
Pearse BMF (1988) Receptors compete for adaptors found in plasma membrane coated pits. EMBO J 7: 3331–3336
Pelham HRB (1988) Evidence that ER luminal proteins are sorted from secreted proteins in a post-ER compartment. EMBO J 7: 913–918
Pfeffer SR (1987) The endosomal concentration of a mannose 6-phosphate receptor is unchanged in the absence of ligand synthesis. J Cell Biol 105: 229–234
Pfeffer SR, Rothman JE (1987) Biosynthetic protein transport and sorting in the endoplasmic reticulum and Golgi. Annu Rev Biochem 56: 829–852
Pfeffer SR, Ullrich A (1986) Structural relationships between growth factor precursors and cell surface receptors. In: Kahn P, Graf T (eds) Oncogenes and growth control. Springer, Berlin Heidelberg, New York, pp 70–76
Pohlmann R, Nagel G, Schmidt B, Stein M, Lorkowski G, Krentler C, Cully J, Meyer HE, Grzeschik K-H, Mersmann G, Hasilik A, Von Figura K (1987) Cloning of a cDNA encoding the human cation-dependent mannose 6-phosphate-specific receptor. Proc Natl Acad Sci USA 84: 5575–5579
Purchio AF, Cooper JA, Brunner AM, Lioubin MN, Gentry LE, Kovacina KS, Roth RA, Marquardt H (1988) Identification of mannose 6-phosphate in two asparagine-linked sugar chains of recombinant transforming growth factor-β1 precursor. J Biol Chem 263: 14211–14215
Quiocho FA (1986) Carbohydrate-binding proteins: tertiary structures and protein-sugar interactions. Annu Rev Biochem 55: 287–315
Rechler MM, Nissley SP (1985) The nature and regulation of the receptors for insulin-like growth factors. Annu Rev Physiol 47: 425–442
Reitman ML, Kornfeld S (1981) Lysosomal enzyme targeting: N-acetylglucosaminyl phosphotransferase selectively phosphorylates native lysosomal enzymes. J Biol Chem 256: 11977–11980
Rose JK, Doms RW (1988) Regulation of protein export from the endoplasmic reticulum. Annu Rev Cell Biol 4: 258–288
Roth RA (1988) Structure of the receptor for insulin-like growth factor II: the puzzle amplified. Science 239: 1269–1271
Roth J, Taatjes DJ, Lucocq JM, Weinstein J, Paulson JC (1985) Demonstration of an extensive trans-tubular network continuous with the Golgi apparatus that may function in glycosylation. Cell 43: 287–295
Roth RA, Stover C, Hari J, Morgan DO, Smith MC, Sara V, Fried VA (1987) Interactions of the receptor for insulin-like growth factor II with mannose 6-phosphate and antibodies to the mannose 6-phosphate receptor. Biochem Biophys Res Commun 149: 600–606
Sahagian GG, (1984) The mannose 6-phosphate receptor: function, biosynthesis, and translocation. Biol Cell 51: 207–214
Sahagian GG, (1987) The mannose 6-phosphate receptor and its role in lysosomal enzyme transport. In: Parent B, Olden K, (eds) Recent research in vertebrate lectins, advanced cell biology monographs. Von Nostrand Reinhold, New York, pp 46-64
Sahagian GG, Neufeld EF (1983) Biosynthesis and turnover of the mannose 6-phosphate receptor in cultured Chinese hamster ovary cells. J Biol Chem 258: 7121–7128
Sahagian GG, Steer CJ (1985) Transmembrane orientation of the mannose 6-phosphate receptor in isolated lathrin-coated vesicles. J Biol Chem 260: 9838–9842d

Sahagian GG, Distler J, Jourdian GW (1981) characterization of a membrane-associated receptor from bovine liver that binds phosphomannosyl residues of bovine testicular β-galactosidase. Proc Natl Acad Sci USA 78: 4289–4293
Salminen A, Novick PJ (1987) A ras-like protein is required for a post-Golgi event in yeast secretion. Cell 49: 527–538
Segev N, Mullholland J, Botstein D (1988) The yeast GTP-binding YPT1 protein and a mammalian counterpart are associated with the secretion machinery. Cell 52: 915–924
Simons K, Virta H (1987) Perforated MDCK cells support intracellular transport. EMBO J 6: 2241–2247
Snider MD, Rogers O (1985) Intracellular movement of cell surface receptors after endocytosis: resialylation of asialo-transferrin receptor in human erythroleukemia cells. J Cell Biol 100: 826–834
Stein M, Meyer HE, Hasilik A, von Figura K (1987a) 46kDa mannose 6-phosphate receptor: purification, subunit composition, chemical modification. Biol Chem Hoppe Seyler 368: 927–936
Stein M, Braulke T, Krentler C, Hasilik A, von Figura K (1987b) 46kDa mannose 6-phosphate-specific receptor: biosynthesis, processing, subcellular location and topology. Biol Chem Hoppe Seyler 368: 937–947
Stein M, Zijderhand-Bleekemolen JE, Geuze H, von Figura K (1987c) M_r 46,000 mannose 6-phosphate specific receptor: its role in targeting of lysosomal enzymes. EMBO J 6: 2677–2681
Tally M, Li CH, Hall K (1987) IGF-2 stimulated growth mediated by the somatomedin type 2 receptor. Biochem Biophys Res Commun 148: 811–816
Todderud G, Carpenter (1988) Presence of mannose phosphate on the epidermal growth factor receptor in A-431 cells. J Biol Chem 263: 17893–17896
Tong PY, Tollefsen SE,Kornfeld S (1988) The cation-independent mannose 6-phosphate receptor binds insulin-like growth factor II. J Biol Chem 263: 2585–2588
Ullrich A, Bell JR, Chen EY, Herrera R, Petruzelli LM, Dull TJ, Gray A, Coussens L, Liao Y-C, Tsubokawa M, Mason A, Seeburg PH, Grunfeld C, Rosen OM, Ramachandran J (1985) Human insulin receptor and its relationship to the tyrosine kinase family of oncogenes. Nature 313: 756–761
Ullrih A, Gray A, Tam AW, Yang-Feng T, Tsubokawa M, Collins C, Henzel W, LeBon, Kathuria S, Chen E, Jacobs S, Franke U, Ramachandran J, Fujita-Yamaguhi Y (1986) Insulin-like growth factor I receptor primary structure: comparison with insulin receptor suggests structural determinants that define functional specificity. EMBO J 5: 2503–2512
van Driel IR, Davis CG, Goldstein JL, Brown MS (1987) Self-association of the low density lipoprotein receptor mediated by the cytoplasmic domain. J Biol Chem 262: 16127–16134
Varki A, Kornfeld S (1980) Structural studies of phosphorylated high-mannose type oligosaccharides. J Biol Chem 255: 10847–10858
Varki A, Kornfeld S (1983) The spectrum of anionic oligosaccharides released by endo-β-N-acetylglucosaminidase H from glycoproteins. J Biol Chem 258: 2808–2818
Varki A, Reitman ML, Kornfeld S (1981) Identification of a variant of mucolipidosis III: a catalytically active N-aetylglucosaminylphosphotransferase that fails to phosphorylate lysosomal enzymes. Pro Natl Acad Sci USA 79: 7773–7777
Viitala J, Carlsson SR, Siebert PD, Fukuda M (1988) Molecular cloning of cDNAs encoding lamp A, a human lysosomal membrane glycoprotein with apparent Mr = 120,000. Proc Natl Acad Sci USA 85: 3743–3747
von Figura K, Hasilik A (1986) Lysosomal enzymes and their receptors. Annu Rev Biochem 55: 167–193
von Figura K, Gieselmann V, Hasilik A (1984) Antibody to mannose 6-phosphate specific receptor induces receptor deficiency in human fibroblasts. EMBO J 3: 1281–1286
von Figura K, Gieselmann V, Hasilik A (1985) Mannose 6-phosphate receptor is a transmembrane protein with a C-terminal extension oriented towards the cytosol. Biochem J 225: 543–547
Waheed A, Pohlmann R, Hasilik A, von Figura K, Van Elsen A, Leroy JG (1982) Deficiency of UDP-*N*-acetylglucosamine: lysosomal enzyme *N*-acetylgluosamine 1-phosphotransferase in organs of I-cell patients. Biothem Biophys Res Commun 105: 1042–1058
Waheed A, Braulke T, Junghans U, von Figura K (1988) Mannose 6-phosphate/insulin-like growth factor II receptor: the two types of ligands bind simultaneously to one receptor at different sites. Biochem Biophys Res Commun 152: 1248–1254
Willingham MC, Pastan IH, Sahagian GG, Jourdian GW, Neufeld EF (1981) Morphologic study of the internalization of a lysosomal enzyme by mannose 6-phosphate receptors in cultured Chinese hamster ovary cells. Proc Natl Acad Sci USA 78: 6967–6971
Yarden Y, Ullrich A (1988) Growth factor receptor tyrosine kinases. Annu Rev Biochem 57: 443–478

Protein Localization and Virus Assembly at Intracellular Membranes

R. F. PETTERSSON

1 Introduction

A large number of viruses contain a lipoprotein coat or envelope that surrounds the internal nucleocapsid or core. This envelope is made up of a regular lipid bilayer derived from and similar in structure and composition to one of the host cell membranes. Viruses acquire their envelope through a budding process,

Ludwig Institute for Cancer Research, Stockholm Branch, Box 60202, S-10401 Stockholm, Sweden whereby the core present in the cytoplasm (or the nucleoplasm, in the case of herpesviruses) pushes itself through specialized regions of a membrane. In this process, host membrane proteins are effectively excluded and replaced by virus-specific membrane proteins. Thus, while virus particles contain only virus-specified proteins, the lipid composition is thought to reflect that of the cellular membrane at which budding has occurred. Enveloped viruses possess one or several integral membrane proteins that form the surface projections (spikes or peplomers). These proteins have important functions, e.g., in mediating attachment to cell surface receptors or in facilitating penetration by triggering fusion of the viral envelope with either the plasma membrane (PM) or the endosomal membrane. Neutralizing antibodies are also directed against these proteins. In some cases, the spike proteins fall into one of three classes of integral membrane proteins (WICKNER and LODISH 1985). The majority are oriented with the N terminus protruding towards the outside and the C terminus facing the core (class I proteins). In some cases (e.g., the neuraminidase, NA, of influenza viruses), the proteins have the opposite orientation (class II), and in yet other cases (e.g., the E1 protein of coronaviruses) the protein may span the lipid bilayer three times (class III). In the first two cases, the bulk of the protein is usually located on the outside of the viral particles (or luminal side of intracellular membranes), while in the latter case the major portion of the protein is facing the nucleocapsid (or cytoplasm).

In many viruses (e.g., orthomyxoviruses, paramyxoviruses, and rhabdoviruses), a matrix or membrane (M) protein is located underneath the lipid bilayer and may participate in shaping the virus structure or contribute to the budding process itself. The spike proteins are synthesized at membrane-bound polysomes in the rough endoplasmic reticulum (ER), where they in most cases become core glycosylated at asparagine residues (*N*-linked glycans). From there, they are transported to the Golgi complex (GC), where *O*-linked glycans may be added to serines or threonines, and further to the PM. Thus, viral glycoproteins destined to travel to the PM follow the general route of exocytosis also utilized by host cell PM proteins (PALADE 1975; KELLY 1985). In fact, viral spike proteins [e.g., the vesicular stomatitis virus (VSV) G protein, the influenza virus hemagglutinin (HA) protein, and the p62 and E1 glycoproteins of the alphaviruses] have been instrumental in dissecting the various steps involved in this exocytic pathway. For example, the importance of glycoprotein oligomerization (trimerization) for the exit from the ER has recently been demonstrated using these viral model systems (COPELAND et al. 1986; KREIS and LODISH 1986; DOMS et al. 1987; COPELAND et al. 1988; EINFELD and HUNTER 1988). However, as will be discussed in this review, not all viral glycoproteins end up at the PM. Instead, they accumulate at one of the internal membranes along the exocytic pathway and thereby determine the site of viral budding.

The components of the core, and the M protein where it exists, are synthesized in the cytoplasm and then assembled and transported to the site of virus budding. The mechanism of virus budding is still largely unknown. It is believed that the assembly process is initiated and progresses by specific

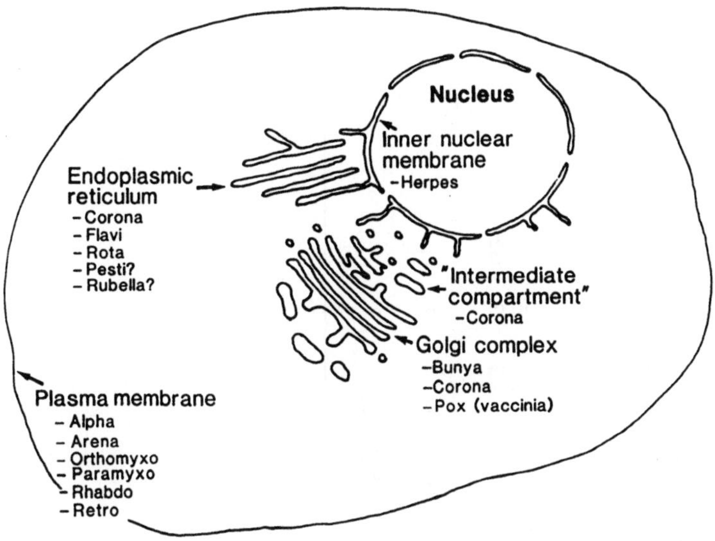

Fig. 1. Cellular membrane compartments utilized for the assembly (budding) of enveloped viruses. Virus families or genera are listed for the different compartments

interactions between the nucleocapsid and a cytoplasmic extension of one (or more) of the viral glycoproteins (SIMONS and GAROFF 1980). Evidence for nucleocapsid-spike interactions has been obtained for Semliki Forest virus (SFV), an alphavirus (GAROFF and SIMONS 1974; HELENIUS and KARTENBECK 1980). Recently, the interaction between the nucleocapsid protein C and the cytoplasmic tail of the E2 glycoprotein of SFV could be demonstrated using a novel approach involving peptide antibodies and antiidiotypic antibodies (VAUX et al. 1988).

Most viruses (e.g., alphaviruses, arenaviruses, orthomyxoviruses, paramyxoviruses, rhabdoviruses, and retroviruses) acquire their envelope at the PM (Fig. 1). After budding has been completed, virus particles are released directly into the extracellular space. However, quite a few viruses are assmbled at intracellular membranes (Table 1, Fig. 1). Such viruses are, in most cases, transported within membrane vesicles to the PM, where they are released from the cell upon fusion of the vesicles with the PM. The purpose of this review is to discuss this latter group of viruses with special reference to the role of the viral glycoproteins in determining the site of virus maturation. It is likely that the accumulation (retention) of one or more of the viral glycoproteins in the compartment where budding occurs plays a decisive role in determining the maturation site. Analyses of the structure and biosynthesis of such organelle- or compartment-specific viral glycoproteins may also be useful as models for studying the molecular basis for the organelle specificity of host cell proteins and for defining "retention signals" responsible for maintaining proteins in defined

Table 1. Assembly of enveloped viruses at intracellular membranes

Membrane compartment	Virus family or genus, examples	Virus envelope glycoproteins	Compartment-specific proteins
Endoplasmic reticulum	Flaviviruses Yellow fever, West Nile	E, M[a]	?
	Rotaviruses Simian SA11	NS28[b], VP7	NS28, VP7
	Non-arthropod-borne and some unclassified togaviruses Pestiviruses, rubella virus	Usually two glyco-proteins	?
Pre-Golgi	Coronaviruses[c] Mouse hepatitis virus, avian infectious bronchitis virus	E1, E2	E1
Golgi complex	Coronaviruses	E1, E2	E1
	Bunyaviruses Hantaan, La Crosse, Punta Toro, Rift Valley fever, Unkuniemi viruses	G1, G2	G1 + G2
	Proxviruses Vaccinia virus	$\geqslant$ Eight on extracellular virus	?
Inner nuclear membrane	Herpesviruses Herpes simplex virus	$\geqslant$ Six (gB, gC, gD, gE, gG, gH)	?

[a] Mature form of M is nonglycosylated
[b] Present only in enveloped particles
[c] May also bud into the endoplasmic reticulum

compartments. In this review, emphasis is put on recent data obtained for the best studied virus models (rotaviruses, coronaviruses, and bunyaviruses). Less well studied virus systems will be discussed only briefly (vacciniaviruses, herpesviruses, and flaviviruses). Various aspects of the assembly of enveloped viruses have been described in some recent reviews and books (SIMONS and GAROFF 1980; DUBOIS-DALCQ et al. 1984; STRAUSS and STRAUSS 1985; SIMONS and FULLER 1987; STEPHENS and COMPANS 1988; HUNTER 1988).

2 Virus Assembly in the Endoplasmic Reticulum

Quite a few viruses are thought to utilize the ER membranes for their maturation. These include the rotaviruses, flaviviruses, and some non-arthropod-borne togaviruses, for example the pestiviruses (bovine diarrhea virus, hog cholera virus), rubellavirus, and some unclassified togaviruses (e.g., equine arteritis virus, lactic dehydrogenase virus) (DUBOIS-DALCQ et al. 1984). Of these, only the rotaviruses, which have been intensively studied in recent years, will be discussed here in detail. For the other viruses, the events leading to virus formation are poorly understood and the final proof that virus budding occurs only or predominantly in the ER is still lacking. Progress in the molecular biology of flaviviruses has been rapid during the past few years, making it now easier to dissect the various steps in virus assembly. Recent advances in the studies of this medically important family are therefore also discussed below.

2.1 Rotaviruses

Members of the *Rotavirus* genus, family *Reoviridae*, are icosahedral, non-enveloped, double-shelled particles that characteristically mature in the rough ER. The genome consists of 11 double-stranded RNA segments each coding for a unique polypeptide. Rotaviruses are a major cause of gastroenteritis in humans and other mammalian species. For details on various aspects of rotaviruses the reader is referred to reviews by ESTES et al. (1983) HOLMES (1983), DUBOIS-DALCQ et al. (1984), KAPIKIAN and CHANOCK (1985), BOTH (1988), BELLAMY and BOTH (1990), and ESTES and COHEN (1990). Much of the information on rotavirus structure and assembly have been elucidated using the simian SA11 rotavirus as a model, since this virus grows to high titers in tissue culture cells.

2.1.1 Morphogenesis and Assembly

Three forms of rotavirus particles are found in infected cells: (1) Single-shelled (ss) particles present in large cytoplasmic inclusions (viroplasms) which are located in close proximity to the ER membranes. (2) Enveloped particles present in the lumen of the ER. These particles have the appearance of regular enveloped viruses with a unit membrane surrounding an inner core. (3) Double-shelled (ds) particles present in the lumen of the ER and lacking the lipid membrane.

The core particles are formed by the condensation of the 11 RNA species with the viral proteins VP1 (125 kDa), VP2 (94 kDa), and VP3 (88 kDa). Some nonstructural proteins (NS53, NS35, and NS34) may also be transiently involved in this process (SOLER et al. 1982). Addition of VP6 (41 kDa), the major inner capsid protein, to the core results in the formation of the ss particles. VP6 possesses an oligomeric (possibly trimeric) structure both in virus particles (GORZIGLIA et al. 1985) and when expressed alone from the cloned cDNA in insect cells using the baculovirus expression system (ESTES et al. 1987). The ss particles aggregate into large, electron-dense inclusions (viroplasms), at the edges of which particles are seen to bud into the lumen of the ER, thus resulting in the transient enveloped form of particles (HOLMES 1983; DUBOIS-DALCQ et al. 1984; PORUCHYNSKY et al. 1985; KAPIKIAN and CHANOCK 1985). As will be discussed below, the nonstructural protein NS28 (NCVP5) may play a critical role in the budding process. How the mature ds particles are formed is still unknown, but somehow the lipid membrane is lost, resulting in the naked mature ds particles that contain, in addition to the proteins present in the ss particles, VP7 (38 kDa) and VP4 (88 kDa) (The name for the outer capsid HA was recently changed from VP3 to VP4: VP3 is the product of segment 3, and VP4 of segment 4; LIU et al. 1988; BURNS et al. 1988.) It is not known at which stage VP4 is incorporated into the particles, prior to or during the budding process. Ultrastructural analysis has localized VP4 to regions between the ER membrane and the viroplasm. In immunofluorescence, VP4 shows a distribution pattern similar to that of NS28 and virion-associated VP7 (KABCENELL et al. 1988). VP7, on the other hand, is probably added to particles after the stripping of the membrane. Enveloped particles accumulate in the ER lumen when *N*-glycosylation is blocked by tunicamycin (SABARA et al. 1982; PETRIE et al. 1983), suggesting a critical role for glycosylation in the stripping of the lipid membrane. Since both VP7 and NS28 are glycoproteins, this effect is probably mediated by either or both of these proteins. The virion-associated VP7 appears to have a conformation different from that of VP7 associated with ER membranes, as revealed by different reactivities with monoclonal antibodies (KABCENELL et al. 1988).

While VP7 is associated with the ER membrane on its luminal side, VP4 seems to be a soluble cytoplasmic protein synthesized on free polysomes. It lacks a classical signal sequence, and potential glycosylation sites are not utilized. VP4 possesses hemagglutinating activity and can influence the extent of viral virulence (KALICA et al. 1983). Proteolytic cleavage of VP4 (to 60-kDA and 28-kDA polypeptides) greatly enhances infectivity (ESPEJO et al. 1981; ESTES et al. 1981). In mature particles VP4 appears to be in close proximity to VP7 (GREENBERG et al. 1983). Both proteins give rise to neutralizing antibodies (OFFIT et al. 1986).

It has been suggested that the ss particles are precursors to the enveloped particles, which in turn are precursors to the ds particles (KABCENELL et al. 1988

AU et al. 1989a). Incorporation of newly synthesized inner capsid proteins into ss particles occurs rapidly (within minutes), while VP4 and VP7 appear in mature ds particles with a lag of 10–15 min. The lag has been suggested to correspond to the time required for virus budding and stripping of the membrane. Such kinetic studies have been facilitated by methods of separating ss from ds particles. The intracellular particles can be purified away from cellular membranes by extraction with trichlorotrifluorocarbon followed by centrifugation. The viral particles are recovered from the aqueous phase while the ER membranes are recovered from the proteinaceous interface. Single- and double-shelled particles can then be separated by CsCl gradient centrifugation due to their different densities (1.355 and 1.335 g/cm³ respectively). So far, pure preparations of enveloped particles have not been obtained (KABCENELL et al. 1988).

Mature virions containing the 11 double-stranded RNA segments and VP1, VP2, VP3, VP4, VP6, and VP7 accumulate in the lumen of the ER and are released from the cell upon cell lysis. No evidence has been presented to indicate that virus particles are transported out of the cell via the exocytic pathway. Thus, some mechanism must exist that causes the virus particles to be retained in the ER. Whether this occurs as the result of virus aggregation or is mediated by one of the viral proteins (e.g., the ER-specific VP7) is not known. Since flaviviruses (see below), for example, are transported out of the ER, the size of the rotavirus particle should not per se preclude its secretion.

2.1.2 NS28 and VP7 are Resident ER Glycoproteins

As for other intracellularly maturing viruses, it is conceivable that one or more of the virus-encoded proteins could determine the site of maturation of rotaviruses in the ER. Two such proteins, NS28 (also known as NCVP5 or VP10) and/or VP7, could serve this function. Both are glycosylated at asparagine residues. Light and electron microscopic studies have indicated that both are confined exclusively to the ER (CHASEY 1980; PETRIE et al. 1982, 1984; PORUCHYNSKY et al. 1985). This conclusion has gained further support from the finding that these proteins contain only high-mannose [endoglycosidase H (endo H)-sensitive] glycans (see below). In CV1 cells, NS28 made from microinjected mRNA was localized by immunofluorescence to the ER and was found to have only endo H-sensitive glycans. A similar result was found after microinjecting the mRNA into *Xenopus* oocytes (ARMSTRONG et al. 1987). The fact that these proteins are not transported out of the ER would thus restrict virus budding exclusively to the ER. While VP7 is found in mature ds particles, NS28 is removed when the enveloped particles are stripped of their membrane.

2.1.2.1 NS28 (NCVP5)

NS28 is encoded by RNA segment 10 and is 175 residues in length, giving a molecular mass (Mr) of about 19000 in the unglycosylated form. The segment 10 RNAs from simian (SA11), bovine (UK), and human (Wa) rotaviruses have been cloned and sequenced (BOTH et al. 1983b; BAYBUTT and McCRAE 1984; OKADA et al. 1984; WARD et al. 1985; POWELL et al. 1988). Comparison of the amino acid sequences indicated a high degree of conservation, with the region between residues 131 and 161 showing the greatest degree of divergence (OKADA et al. 1984; WART et al. 1985; POWELL et al. 1988). There are two potential sites for *N*-glycosylation near the N terminus of NS28 (at residues 8 and 18). They are conserved among all examined rotaviruses and appear both to be used, resulting in an Mr of 28000 for the fully glycosylated form. In conformity with their localization to the ER, the glycans are of the high-mannose type (ERICSON et al. 1982; BOTH et al. 1983b) and are trimmed during a 60-min chase only to the $Man_9GlcNAc_2$ and $Man_8GlcNAc_2$ stages (KABCENELL and ATKINSON 1985), with the former being the predominant species.

Several lines of evidence indicate that NS28 is a transmembrane protein with the N terminus located on the luminal side and the C terminus on the cytoplasmic side (Fig. 2). Firstly, the two glycosylation sites are used and must be on the luminal side of the ER membrane. Secondly, the cytoplasmic portion is susceptible to protease treatment. Treatment of microsomes from infected cells, or of membranes into which NS28 has been inserted in an in vitro translation system with, for example, trypsin, chymotrypsin, or proteinase K, substantially reduces the size of NS28. Taken together, these protection studies indicate that about 130 amino acids are located on the cytoplasmic side of the ER membrane (ERICSON et al. 1983; KABCENELL and ATKINSON 1985; CHAN et al. 1988; BERGMANN et al. 1989). NS28 thus resembles cytochrome p450, for example, in its topology in the ER membrane (SAKAGUCHI et al. 1987).

The deduced amino acid sequence of NS28 predicts the presence of three hydrophobic domains located in the N-terminal half of the protein. They map at residues 7–21 (H1), 28–47 (H2), and 67–85 (H3). The two glycosylation sites are located within H1 (BAYBUTT and McCRAE 1984; POWELL et al. 1988; CHAN et al. 1988; BERGMANN et al. 1989), indicating that H1 is not cleaved and is luminally located. Based on protease protection studies and by using a hydrophobic

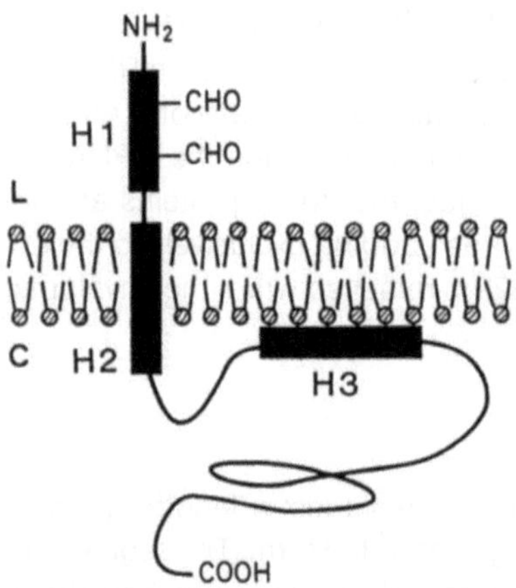

Fig. 2. Proposed topology of the rotavirus NS28 glycoprotein in the ER membrane. *H1-H3*, the three hydrophobic transmembrane domains; *CHO*, *N*-linked glycans; *L*, lumen; *C*, cytoplasm. (Adapted from BERGMANN et al. 1989)

moment plot to analyze potential transmembrane regions, CHAN et al. (1988) arrived at a model suggesting that H3 is the membrane-spanning domain, while H1 is partly and H2 completely buried in the lipid bilayer. However, by expressing deletion mutant of NS28 in an in vitro transcription-translation system, BERGMANN et al. (1989) arrived at a different conclusion. According to their results, H2 serves both as the signal sequence and as the membrane-spanning "stop transfer" sequence, while H3 is embedded in the lipid bilayer on the cytoplasmic side of the membrane (Fig. 2). These conclusions were further strengthened by protease protection studies. A trypsin- and proteinase K-sensitive region between H2 and H3 was found on the cytoplasmic side of the membrane. The H1 domain could not mediate translocation through the ER membrane. Deletion analysis indicated that residues 28–40 within H2 are sufficient for mediating membrane translocation. A mutant lacking H1 and H2 was also able to insert into membranes, suggesting that H3 has the ability to serve as a signal sequence, although it may not have that function in the normal protein.

To date, the amino acid sequence(s) responsible for the retention of NS28 in the ER has not been identified. Both wild type and various deletion mutant forms of NS28 expressed from cloned cDNA localized to the ER and acquired only endo H-sensitive glycans (ARMSTRONG et al. 1987; BERGMANN et al. 1989). NS28, being a small and well-characterized protein, should, however, be an ideal model for defining an ER retention signal.

Recently, it has been shown that NS28 serves as the receptor for binding of ss particles to the ER membrane. Microsomal membranes isolated from virus-infected cells, or from insect cells expressing the NS28 protein from a baculovirus expression vector, were found to specifically bind CsCl gradient-purified ss particles (AU et al. 1989a, b). MAYER et al. (1989) likewise found that microsomal membranes isolated from cells infected with a vaccinia recombinant virus expressing NS28 could bind core particles. The binding was stimulated by Ca^{2+} and Mg^{2+} and could be inhibited by a monoclonal antibody against VP6, or by purified VP6 protein. AU et al. (1989b) also found that VP6 seems to mediate binding to NS28. With these in vitro systems at hand it should now be possible to identify the sequences at the C-terminal end of NS28 responsible for binding and to characterize the conditions required for both binding and budding. For instance, calcium has been shown to be critical for virus morphogenesis. Depletion of calcium from the medium of virus-infected cells results in the accumulation of ss particles and degradation of VP7 (SHAHRABADI et al. 1987). Since the prerequisites for virus budding have not yet been defined for any virus, the rotaviruses offer a convenient model system for studying this crucial event in virus assembly.

2.1.2.2 VP7

VP7 is the product of RNA segment 9, the nucleotide sequence of which has been determined for several strains (BOTH et al. 1983a; RICHARDSON et al. 1984; GLASS et al. 1985; GUNN et al. 1985; GORZIGLIA et al. 1986). VP7 is the serotype-specific neutralizing antigen with a structure that is highly conserved among rotaviruses (RICHARDSON et al. 1984; GLASS et al. 1985; GUNN et al. 1985; ANDREW et al. 1987). The protein is *N*-glycosylated and associated with ER membranes (ERICSON et al. 1983; KABCENELL and ATKINSON 1985). Since there are two potential in-frame AUG initiation condons in the gene (corresponding to amino acid residues 1 and 30), the size of the primary translation product has been unclear. If initiation occurs at the first AUG, the protein would be 326 residues in length. The two methionine codons are each followed by a hydrophobic stretch of amino acids (H1, between residues 6 and 23; H2, between residues 32 and 48), both of which could potentially serve a signal peptide function. Recent results obtained by expressing in vitro mutagenized VP7 cDNA, as well as by determination of the N-terminal amino acid sequence of VP7 made in vitro in the presence of microsomal membranes, have indicated that the major site of initiation is the second AUG (corresponding to residue 30 according to the original numbering). The primary translation product is then cleaved probably by the signal peptidase between Ala_{50} and Gln_{51}. Thus, the final product is only 276 residues in length (STIRZAKER et al. 1987; WHITFIELD et al. 1987; PORUCHYNSKY and ATKINSON 1988). The first AUG codon is in a "weak" context according to the Kozak consensus rule (KOZAK 1986). Whether this AUG is used in vivo is still a matter of debate. It is noteworthy that the two AUG condons as well as the H1 and H2 domains have been conserved among rotaviruses, thus suggesting important functions for both domains. While H2 seems to serve as the signal peptide, in vivo H1 can substitute for that function, since a mutant lacking the H2 region was still able to anchor VP7 in the ER membrane (WHITFIELD et al. 1987). CHAN et al. (1986) have described the presence of two forms of VP7 in infected cells and suggested a processing between H1 and H2. These results have not been corroborated by the recent work of STIRZAKER et al. (1987).

VP7 was originally suggested to be an integral membrane protein anchored to the ER membrane by the second N-terminal hydrophobic domain (KABCENELL and ATKINSON 1985). Although VP7 clearly seems to be membrane-associated, the recent finding that a cleavage occurs after Ala_{50} means that the hydrophobic regions H1 and H2 are both removed. The question therefore arises of how the now hydrophilic VP7 can still remain membrane-bound and be confined to the ER. One possibility would be that VP7 binds to a cellular ER-specific protein. The fact that VP7 when transfected alone into COS cells is retained in and anchored to the ER membrane argues against a role for any other viral proteins in the ER retention. Originally, mutants deleted in the H2 domain, between residues 42 and 61 suggested that the signal for ER retention might be located in the H2 domain (PORUCHYNSKY et al. 1985). Reexamination of these mutants as well as analyses of further mutants of the N-terminal region of VP7 have, however, indicated that two regions are involved in retaining VP7 in the ER. These mapped between residues 51 and 61, and between 61 and 111. Neither of these regions is alone sufficient to keep VP7 in the ER, nor to cause the retention of mouse salivary α-amylase in the ER. However, when the first 111 residues were added onto the N terminus of amylase from which its own signal sequence had been deleted, amylase was translocated and retained in an enzymatically active form in the ER. No amylase was secreted. Thus, the ER retention signal was transferable to a protein normally secreted out of the cell (PORUCHYNSKY and ATKINSON 1988).

The question of the nature of the retention signal in VP7 took an unexpected and exciting turn when STIRZAKER and BOTH (1989) recently were able to show that the H2 signal peptide is essential for ER retention. When the H2 domain was replaced by the signal peptide from an influenza virus HA, normal translocation and processing of VP7 occurred, but VP7 was readily secreted. The H2 peptide, when transferred to a heterologous protein, the malaria S antigen, was not able to hold back the protein in the ER. Taking these and previous results together it thus appears that the retention of VP7 in the ER involves an interaction between the H2 signal peptide and some downstream region(s) within the first 111 residues of VP7. It will be of great interest to define the minimal sequence required for ER retention. It is noteworthy that the C-terminal Lys-Asp-Glu-Leu sequence present in several ER-specific luminal proteins (MUNRO and PELHAM 1987; PELHAM 1988; GEETHA-HABIB et al. 1988) and responsible for their retention in the ER is not present in VP7.

The single *N*-linked glycan of VP7 (at residue 69) is processed during a 90-min chase predominantly to $Man_8GluNAc_2$ and Man_6GluAc_2. A minor fraction of the glycans in VP7 in mature ds particles is, in addition, processed down to $Man_5GluNAc_2$ (ERICSON et al. 1983; KABCENELL and ATKINSON 1985; KABCENELL et al. 1988). Since VP7 is thought not to leave the ER and virus particles have not been seen in the Golgi complex, it is possible that the removal of the mannose residues occurs as the result of action of ER mannosidases. Since NS28 is trimmed to a lesser degree (see above) than VP7, this could mean that VP7 and NS28 become exposed to different mannosidases, perhaps in different subcompartments of the ER. The alternative possibility that VP7 is transported to the cis-Golgi and then recycled back to the ER (ROTHMAN 1981) cannot, however, be ruled out at present (KABCENELL and ATKINSON 1985). VP7 thus offers a good model also for studying the trimming of glycans within the ER.

The role of VP7 in the assembly of rotavirus particles still remains to be elucidated. Are there interactions between VP7, NS28, and VP4 during the budding process? When and how is VP7 added to virus particles? Does this process resemble the transfer of the membrane-associated coat protein of the filamentous phage M13 to mature particles (WEBSTER and CASHMAN 1978).

In summary, the rotaviruses mature by budding into the lumen of the ER, where the lipid membrane is removed. This site of maturation may be determined by the rotavirus glycoprotein NS28 (AU et al. 1989a, b; MEYER et al. 1989), which has been shown to be a resident ER protein. This protein and the ER-specific virion protein VP7 are good models for studying the targeting of proteins to the ER and for analyzing the processing of glycans within the ER compartment. Because nucleocapsids can be bound to ER membranes harboring NS28, this system should also be ideal to study the budding event in vitro.

2.2 Flaviviruses

The flaviviruses are small (diameter 40–50 nm) enveloped particles that acquire their envelope in the ER. The single-stranded RNA genome has a positive polarity and is about 10 500 nucleotides in length. Most of the members of the Flaviviridae family are spread by arthropods, and many are important human pathogens, causing encephalitides, hemorrhagic fevers, arthralgias, and rash (e.g., tick-borne encephalitis, Japanese encephalitis, St. Louis encephalitis, West Nile, yellow fever, dengue viruses). A number of reviews have appeared covering different aspects of flavivirus structure and replication (MURPHY 1980; MONATH 1985; RICE et al. 1986; DUBOIS-DALCQ et al. 1984; BRINTON 1986; WESTAWAY 1987). Since the assembly process of flaviviruses is rather poorly understood, this group of viruses will be discussed only briefly here. The morphogenesis of flaviviruses has been critically reviewed by MURPHY (1980).

2.2.1 Morphogenesis and Assembly

The most prominent feature of flavivirus-infected cells is the proliferation and hypertrophy of both smooth and rough ER membranes. These changes begin in a perinuclear region and expand towards the periphery as infection progresses (MURPHY. 1980; DUBIOS-DALCQ et al. 1984). Although large amounts of flavivirus particles accumulate in the ER during infection, the site of virus maturation has remained somewhat controversial. In general, the unambiguous proof of virus maturation at any particular site is the demonstration by electron microscopy of budding virus particle profiles. For flaviviruses, it has been extremely difficult to find such profiles in which the unit membrane of the ER and the virus envelope are contiguous. Only in rare instances have budding particles been observed (MATSUMURA et al. 1977; MURPHY 1980; ISHAK et al. 1988), and some of these observations are not wholly convincing. Thus, it is possible that some particles are formed at the PM or even at some other membranes, although the bulk clearly appears to be formed at the ER membrane. The absence of budding profiles suggests that this last step is very rapid, making it difficult to catch a forming particle by electron microscopy. Another problem has been the difficulty in distinguishing viral nucleocapsids from ribosomes. Free nucleocapsids are not seen, and the formation of nucleocapsids and the budding process itself appear to be closely linked (MURPHY 1980). As discussed below, nucleocapsid assembly may take place at the ER membrane.

The complete nucleotide sequences of the RNA genome of several flaviviruses have been reported recently, including yellow fever (YF) (RICE et al. 1985), West Nile (WN) (CASTLE et al. 1985, 1986), Japanese encephalitis (JE) (SUMIYOSHI et al. 1987), dengue 2 (DEN) (HAHN et al. 1988), and Kunjin (KUN) (COIA et al. 1988). These sequences show that all virus-encoded proteins are translated as a polyprotein, which is proteolytically cleaved to yield the final polypeptides. Purified virions contain only three structural proteins: the nucleocapsid protein C (13 kDa) and the two membrane proteins E (50 kDa) and

M (8 kDa). The genes encoding these proteins are located in the N-terminal portion of the precursor in the order NH_2-C-preM-E-COOH. The E protein is a typical glycosylated integral transmembrane protein and forms the viral spikes (peplomers). It has hemagglutinating activity and probably mediates the attachment of the virus to cell surface receptors. In virions, the E protein seems to be present as trimers (WENGLER and WENGLER 1989). Most of the M protein is located on the inside of the viral membrane and thus resembles a matrix protein typical for orthomyxoviruses, paramyxoviruses, and rhabdoviruses. After cleavage from the primary translation product, the M protein first appears in a precursor form called preM (22 kDa) that contains *N*-linked glycans and is incorporated into virions. Prior to release of virions from the cell, the N-terminal presequence is cleaved off, generating the mature unglycosylated M protein. The presequence is not found in the virion. In immature intracellular particles, preM and E appear to form heterodimers (WENGLER and WENGLER 1989), which dissociate after the cleavage of preM and release of virions from the cell.

As shown recently for WN virus, the C protein also appears first when translated in vitro in a slightly larger form than mature C, containing an 18-residue-long C-terminal extension (NOWAK et al. 1989). This sequence is very hydrophobic and serves as the signal peptide for the downstream preM protein. It has been suggested that the initial cleavage between the C protein and the preM protein is carried out by the signal peptidase. Prior to or during virus assembly the anchored C protein is converted into mature C protein by an as yet unknown protease that removes the hydrophobic sequence. The presence of the hydrophobic extension suggests that newly made C protein is membrane-bound, a hypothesis which has recently gained support from in vitro translation experiments carried out in the presence of microsomal membrane (NOWAK et al. 1989). In addition, like the membrane proteins, intracellular C protein is only found membrane-associated (SHAPIRO et al. 1972; STOHLMAN et al. 1975; BOULTON and WESTAWAY 1976). Thus, it is possible that the nucleocapsids are assembled in association with the ER membrane followed by cleavage of the C terminus of the C protein and budding of the nucleocapsid through the ER membrane (WENGLER and WENGLER 1989). This is supported by the fact that no free nucleocapsids are found in the cytoplasm of infected cells.

Analyses of the distribution of the membrane proteins in JE and dengue virus-infected cells by subcellular fractionation have indicated that the proteins are present in all cellular membranes, including the PM (SHAPIRO et al. 1972; STOHLMANN et al. 1975; BOULTON and WESTAWAY 1976). According to immunofluorescence studies of YF virus-infected cells, the E protein is first detected around the nucleus and later also in the rest of the cytoplasm. The data available as yet are not sufficiently clear to permit any conclusions as to whether any of the membrane proteins of flaviviruses (or the non-arthropod-borne togaviruses) accumulates in the ER and thereby determines the site of virus assembly. One intriguing possibility is that the site of budding in the case of these viruses may be determined by the assembly of the nucleocapsids at the ER membrane (see above) (NOWAK et al. 1989).

3 Virus Assembly in the Endoplasmic Reticulum, a Pre-Golgi Compartment, or the Golgi Complex

3.1 Coronaviruses

3.1.1 Morphogenesis and Assembly

Members of the Coronaviridae family are enveloped RNA viruses that acquire their lipoprotein coat by budding at intracellular membranes (the ER, the GC, or in a compartment between these two organelles). The single-stranded RNA genome of the coronaviridae type species avian infections bronchitis virus (IBV) is 27 608 residues in length (BOURSNELL et al. 1987) and has a positive polarity. Viruses of this family cause respiratory and/or enteric infections in humans and domestic animals. The models most widely used to study the molecular biology of coronaviruses include the A59 and JHM strains of the mouse hepatitis virus (MHV-A59 and MHV-JHM) and IBV. Details on various aspects of coronaviruses can be found in reviews by SIDDELL et al. (1983), DUBOIS-DALCQ et al. (1984), HOLMES (1985), STURMAN and HOLMES (1983, 1985), and SPAAN et al. (1988).

The coronaviruses have a simple protein composition. The N protein (59 kDa), which in most cases is encoded by the smallest mRNA species, associates in the cytoplasm with the genomic RNA to form the helical, loosely coiled nucleocapsid. All coronaviruses contain two envelope glycoproteins, E1 (also known as M; 23 kDa) and E2 (or S; 180 kDa). The molar ratio of N, E1, and E2 in purified virions has been reported to be 8:16:1 (STURMAN et al. 1980). With a few exceptions (e.g., IBV, transmissible gastroenteritis virus, some strains of MHV), coronaviruses have recently been found to contain a third membrane glycoprotein, E3 (60–65 kDa). This protein contains *N*-linked glycans, occurs as a disulfide-linked homodimer, and exhibits receptor-binding, hemagglutinating, and receptor-destroying (acetylesterase) activities. The latter is important for infectivity (DEREGT et al. 1987; VLASAK et al. 1988).

The N protein is phosphorylated at threonine and serine residues (STOHLMAN and LAI 1979), but the role of this phosphorylation in virus assembly and replication is unknown. During infection, viral nucleocapsids accumulate in the cytoplasm and only a small fraction can be chased into virions (ROTTIER et al. 1981). Late in the infection large cytoplasmic nucleocapsid inclusions can be seen (CAUL and EGGLESTONE 1977). On immunofluorescence, patches of N protein are observed throughout the cytoplasm early in the infection. Such foci increase in number and size as infection progresses.

Only E1 is required for virus budding (see below), since virus particles containing only the nucleocapsid and E1, but no E2, can be assembled in tunicamycin-treated cells (HOLMES et al. 1981; STERN and SEFTON 1982).

Coronavirus budding occurs at intracellular membranes. Nucleocapsids align on the cytoplasmic side of smooth membranes in the ER, or the GC at sites where viral glycoproteins have accumulated (MASSALSKI et al. 1982). Budding at the PM has not been observed. Detailed electron microscopical studies of MHV-A59 virus-infected cells have been carried out to determine the exact site of virus maturation (TOOZE et al. 1984, 1987, 1988; TOOZE and TOOZE 1985). In sac⁻ cells (Moloney sarcoma virus-transformed murine fibroblasts), the first site of virus budding early in the infection occurs at smooth perinuclear vesicles and tubules in a pre-Golgi region located between the ER and the GC. At later stages of infection, budding also occurs in the ER, where virus particles accumulate. Only rarely was the virus seen to bud into the lumen of the Golgi cisternae in this cell line (TOOZE et al. 1984). Between 5% and 10% of AtT20 cells (a murine pituitary tumor cell line) are susceptible to MHV-A59 virus infection. In these cells, virus budding is restricted to the GC, which, during the course of infection, gradually vesiculates (TOOZE and TOOZE 1985; TOOZE et al. 1987). Thus, the site of coronavirus budding is at least partly determined by the host cell.

The nature and function of the pre-Golgi compartment in sac⁻ cells into which coronavirus particles bud is not clear, but this compartment is of great potential interest. By analyzing the kinetics of *O*-glycosylation of E1 (see below), TOOZE et al. (1988) concluded that this budding compartment, which from morphological studies (TOOZE et al. 1984) seems to lie between the ER and the GC, is the site for the additon of *N*-acetylgalactosamine, the first *O*-linked sugar residue added to E1. As will be discussed below, E1 accumulates in this budding compartment and in the GC and does not reach the PM. E2, on the other hand, is transported to the PM via the GC, where, together with E2, present in cell-associated extracellular virions, it induces cell-cell fusion. Where and how E2 is incorporated into virions is not known. One likely possibility is that E2 is included in the budding virus while it is being transported to the PM via the budding compartment. Virions in the ER seem to contain the large club-shaped spikes (peplomers) characteristic of coronaviruses (CHASEY and ALEXANDER 1976). These are made up of E2. After budding, virus particles are transported via the GC in transport vesicles that eventually fuse with the PM (DOUGHRI et al. 1976). Large amounts of released virus particles remain cell-associated (SUGIYAMA and AMANO 1981; DUBOIS-DALCQ et al. 1984). In AtT20 cells, these vesicles follow the constitutive exocytic pathway (TOOZE et al. 1987). Proteins are secreted out of these cells via both a constitutive and a regulated route (KELLY 1985). For instance, ACTH, a cleavage product of the pro-opiomelanocortin precursor, is secreted via the latter route. Coronavirus and secretory proteins traverse the same Golgi cisternae and pass through the trans-Golgi network together. Thereafter their pathways diverge: secretory proteins condense and move to the secretory granules, while vesicles containing coronavirus particles and devoid of secretory proteins are transported directly to the plasma membrane (TOOZE et al. 1987).

3.1.2 E1 Determines the Site of Coronavirus Maturation

The E1 protein is synthesized from mRNA number 6 and is 225 (IBV) or 228 (MHV) amino acids in length, giving an Mr of about 26 000. The sequences of cDNAs encoding E1 from MHV-A59 and IBV have been determined (ARMSTRONG et al. 1984; BOURSNELL et al. 1987) and found to display 25% identity at the amino acid level. The hydropathic pattern is, however, very well conserved. In vitro translation in the presence of microsomal membranes has shown that 60%–70% (140–150 residues) of E1 is synthesized before insertion into the microsomal membrane occurs (ROTTIER et al. 1984). This suggests an internal signal sequence rather than an N-terminal one. Translocation through the microsomal membrane is signal recognizing particle (SRP)-dependent (ROTTIER et al. 1985) and does not involve a cleavage of the signal sequence (ROTTIER et al. 1984; MAYER et al. 1988). The hydropathic plot of E1 reveals one short (residues 25–47) and one long (residues 57–106) hydrophobic domain located within the N-terminal half (ARMSTRONG et al. 1984; MAYER et al. 1988). The latter domain is long enough to span the lipid bilayer twice and can be divided into two hydrophobic domains. Based on the primary structure and on protease protection studies, it has been concluded that E1 spans the membrane three times, leaving a short 22 (IBV) or 25 (MHV)-residue-long glycosylated N-terminal region protruding into the lumen and a ca. 124-residue-long cytoplasmic domain (Fig. 3) (ARMSTRONG et al. 1984; ROTTIER et al. 1984, 1986; MAYER et al. 1988). Thus, E1 can be classified as a class III membrane glycoprotein (WICKNER and LODISH 1985). In vitro mutagenesis combined with both in vitro and in vivo expression of deleted forms of E1 has shown that the first and third hydrophobic domains both can serve as signal sequences (MACHAMER and ROSE 1987; MAYER et al. 1988). Which of the two domains serves this function normally is not known. Because of its topology, it has been proposed that E1, in addition to being a transmembrane

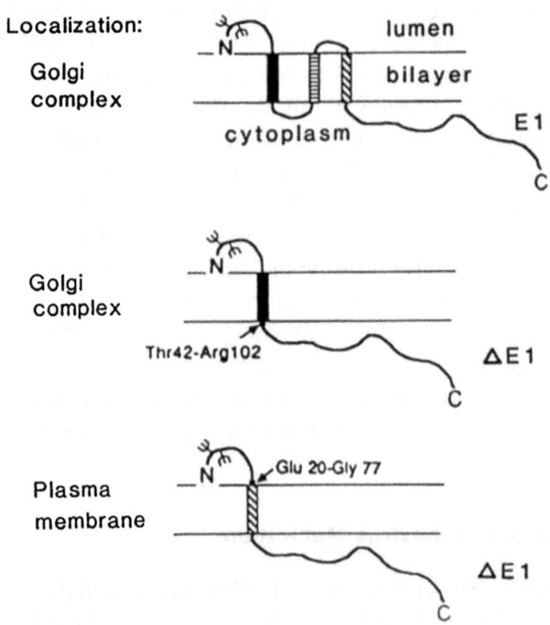

Fig. 3. Proposed topology of the coronavirus (avian infectious bronchitis virus), E1 glycoprotein in the Golgi membrane (*top*). Also shown are the cellular locations of two deletion mutants lacking the second and third (*middle*) or the first and second (*bottom*) hydrophobic transmembrane domains. The amino acid residues fused in the mutants are indicated, as are the two sites for *N*-glycosylation (Adapted from MACHAMER and ROSE 1987)

protein, also may function as a matrix protein (STURMAN and HOLMES 1983) similar to the M protein of orthomyxoviruses, paramyxoviruses, and rhabdoviruses. The cytoplasmic domain of E1 is largely embedded in the membrane, since only a 1.5- to 2-kDa hydrophilic fragment is removed from the C terminus after proteinase K treatment of microsomes (ROTTIER et al. 1984; MAYER et al. 1988). Since only E1 is required for virus budding (HOLMES et al. 1981), a specific interaction between the cytoplasmic domain of E1 and the nucleocapsid must exist. Such an interaction between E1 and nucleocapsids has been observed in vitro (STURMAN et al. 1980).

E1 in MHV contains two to three *O*-linked glycans (NIEMANN and KLENK 1981; NIEMANN et al. 1982, 1984), while IBV contains two *N*-linked glycans (STERN and SEFTON 1982). The *O*-linked glycans are attached to the threonine and serine residues in the sequence NH_2-Met-Ser-Ser-Thr-Thr at the N-terminal end of E1 in MHV. In purified virions, 35% of the glycans have the structure GalNAc-Gal-NeuNAc, while 65% have the branched structure GalNAc(-NeuNAc)-Gal-NeuNAc and exhibit blood group M activitiy (NIEMANN et al. 1984). Pulse-chase experiments in combination with the use of reduced temperature have shown that the *O*-glycosylation occurs in two discrete steps. The first residue, *N*-acetyl-galactosamine, is added within 10 min after the pulse and before E1 reaches the GC, probably in a pre-Golgi compartment that may correspond to the budding compartment (see above). The galactose and sialic acid residues are added some 10 min later, probably in the trans-Golgi (TOOZE et al. 1988).

Interestingly, the *O*-glycosylation of E1 can be inhibited by monensin, an ionophore that blocks the transport of most exported proteins at the level of the medial-Golgi (GRIFFITHS et al. 1983). While this result is consistent with the conclusion that the second *O*-glycosylation step takes place in the trans-Golgi, the inhibition of the first step is more difficult to explain (NIEMANN et al. 1982).

In E1 of IBV, core glycosylation occurs at the Asn_3 and Asn_6 residues. The glycans are then trimmed during transport to the GC into endo H-sensitive and endo H-resistant forms. In virions, both endo H-sensitive and -resistant forms are present (STERN and SEFTON 1982).

Light- and electron-microscopic immunolocalization of E1 in coronavirus-infected cells or cells transfected with a cloned cDNA have indicated that it is predominantly confined to smooth membranes of the pre-Golgi region and the GC and, later in infection, also to the ER (ROTTIER and ROSE 1987; MACHAMER and ROSE 1987; TOOZE et al. 1984, 1988; MAYER et al. 1988). The site of virus budding correlates with the distribution of E1 (TOOZE et al. 1984). Microinjection of E1 mRNA into *Xenopus* oocytes (ARMSTRONG et al. 1987), CV-1 cells, or AtT20 cells (MAYER et al. 1988) also results in the intracellular accumulation of E1. E1 is not transported beyond the trans-Golgi. Thus, the sequence information present in E1 is alone sufficient for targeting to the GC. In IBV-infected cells the *N*-linked glycans of E1 are partly processed to endo H resistance, and E1 in virus particles also possesses some terminally glycosylated glycans (STERN and SEFTON 1982). In MHV-infected cells the *O*-linked glycans are sialylated, a process that is considered to occur in the trans-Golgi. Released particles also contain sialylated glycans (see above). These results could mean that terminal glycosylation of E1 occurs on virus particles while they are being transported through the GC. Alternatively, E1 could first be transported to the trans-Golgi and then recycled back to the budding site. Such recycling has not, however, been demonstrated.

Attempts have been made to localize a "Golgi retention signal" that would be responsible for keeping E1 in that compartment. By expressing deletion mutants of E1 of IBV, MACHAMER and ROSE (1987) concluded that a mutant E1 containing only the first (N-terminal) transmembrane domain localized to the GC, while a mutant E1 with only the third transmembrane domain was transported to the cell surface (Fig. 3). This suggests that the information specifying accumulation in the GC resides in the first transmembrane domain. In conformity with these results, the IBV E1 cytoplasmic tail, when replacing the VSV G cytoplasmic tail, did not block the surface expression of the G protein (PUDDINGTON et al. 1986; MACHAMER and ROSE 1987). MAYER et al. (1988), using deletion mutants of MHV-A59 E1, excluded the possibility that the first 28 N-terminal hydrophilic residues had a targeting function. Similar deletion experiments with E1 of the A59 virus have, in contrast, suggested that a targeting signal may reside in the last 25 C-terminal amino acids (J. Armstrong, personal communication). Deletion of the first and second or of the second and third transmembrane domains did not affect the Golgi targeting. One possible interpretation of these seemingly conflicting results is that cooperation between the first transmembrane domain and the C-terminal end is needed for proper targeting to the GC. Further work has to be done to explore this possibility.

3.1.3 The E2 Glycoprotein

The E2 protein of coronaviruses forms the spikes or peplomers clearly visible on virus particles. The cDNA sequence (derived from mRNA 3) has been determined for E2 of IBV (1162 residues: BINNS et al. 1985) and of MHV strain JMH (1235 residues; SCHMIDT et al. 1987) and strain A59 (1324 residues; LUYTIES et al. 1987). There are 28 and 21 potential sites for *N*-glycosylation in the IBV and MHV-JMH E2 sequences respectively. The apoprotein with an Mr of about 120000 is core-glycosylated in the ER and then processed to a 180-kDa form. During transport, E2 is acylated and at a later stage proteolytically cleaved into two 90-kDa products, S1 (90B) and S2 (90A) (SCHMIDT et al. 1987; NIEMANN et al. 1982; STURMAN et al. 1985). Cleavage probably occurs at the sequence Arg-Arg-Ala-Arg (residues 624-628 in MHV-JMH). Only S2 is acylated, and it mediates the attachment of S1 to the membrane (CAVANAGH 1983; STURMAN et al. 1985). S2 is anchored to the membrane via a 34-residue-long hydrophobic domain located at the C terminus. Only 27 amino acids are exposed on the cytoplasmic side of the membrane (BINNS et al. 1985; SCHMIDT et al. 1987).

Pulse-chase experiments indicate that most of E2 is incorporated into virions and released from the cell within 2 h (HOLMES et al. 1981). As discussed above, E2 is apparently incorporated into virions in the budding compartment while en route to the plasma membrane. Tunicamycin inhibits core glycosylation, but not acylation, of E2 and its transport out of the ER. This results in virus particles containing only the nucleocapsid and E1. Thus, E2 is not required for virus budding. Such particles are, however, noninfectious (HOLMES et al. 1981; STERN and SEFTON 1982; REPP et al. 1985). Inhibitors of the ER glycosidases I and II inhibit MHV E2 transport out of the ER and also drastically reduce the synthesis of E1, resulting in a substantial drop in virus formation. Inhibitors of mannosidases I and II, on the other hand, have little or no effect on the transport of E2 or particle formation (REPP et al. 1985).

E2 mediates the binding of coronavirus to its specific receptor (BOYLE et al. 1987), induces neutralizing antibodies, mediates hemagglutination and possesses fusion activity (for references see, e.g., DUBOIS-DALCQ et al. 1984; HOLMES 1985). For IBV and MHV, cleavage of E2 is required for cell fusion activity by exogenous virions. For other strains, cleavage does not seem to be necessary for fusion activity. Cleavage does not result in the exposure of a highly hydrophobic N-terminal "fusogenic" sequence on the C-terminal fragment (S2) (SCHMIDT et al. 1987) as in the case of orthomyxoviruses or paramyxoviruses. Thus, the mechanism of coronavirus-induced fusion is still unknown. Most of the fusion activity is caused by virus particles associated with the plasma membrane. Uncleaved and fusion incompetent E2 can be activated by trypsin (STURMAN and HOLMES 1985).

In summary, the two virion glycoproteins of coronaviruses have quite different properties. The E1 glycoprotein (a) is inserted into the ER membrane with the aid of an internal and noncleavable signal sequence, (b) spans the membrane three times and has the properties of a typical viral matrix protein, (c) serves as the "receptor" for binding of the nucleocapsid to the membrane, thereby facilitating virus budding, and (d) accumulates in a pre-Golgi compartment and the GC and thus determines the site of virus maturation. The budding site may define a new compartment located between the ER and the GC along the exocytic pathway. The E2 glycoprotein, on the other hand, (a) is a transmembrane protein with a typical N-terminal cleavable signal sequence and a short cytoplasmic tail (the bulk of the protein protrudes out from the virion and constitutes the club-like spike or peplomer), (b) is readily transported to the PM, and (c) has several important functions, including attachment to the cell-surface receptor and induction of cell-cell fusion. Thus, both proteins are good models for studying a variety of cell biological problems related to intracellular protein targeting, transport, and processing.

3.2 Bunyaviruses

Members of the Bunyaviridae family are collectively called bunyaviruses although they fall into five serologically distinct genera: *Bunyavirus, Hantavirus, Nairovirus, Phlebovirus,* and *Uukuvirus* (BISHOP 1985; GONZALEZ-SCARANO and NATHANSON 1990; SCHMALJOHN and PATTERSON 1990). Some 250 different bunyaviruses have been classified so far. Some of them are of veterinary or medical importance, and may cause encephalitides and hemorrhagic fevers in man. Common to all bunyaviruses is a similar general structure and the site of maturation in the GC. This latter characteristic was originally used as a criterion for the creation of the Bunyaviridae family (MURPHY et al. 1973). Only a few reviews have appeared on the structure and biology of bunyaviruses (BISHOP and SHOPE 1979; BISHOP 1985; PETTERSSON and VON BONSDORFF 1987; PETTERSSON et al. 1988; SCHMALJOHN and PATTERSON 1990; GONZALEZ-SCARANO and NATHANSON 1990).

3.2.1 Morphogenesis and Assembly

Despite considerable variation in the Mr of the structural proteins and the genomic RNA segments, the overall structure of the various bunyaviruses is very similar. Virus particles measure about 90–100 nm in diameter and contain four proteins: two glycoproteins, G1 and G2, associated with the envelope and forming the spikes, and two internal proteins associated with the RNA genome, the major nucleocapsid protein N, and the minor RNA-dependent RNA polymerase L. The proteins are encoded by three single-stranded, circular RNA segments, called L (large), M (medium), and S (small), of negative polarity. The L protein is encoded by the L RNA; G1 and G2 (and in some species a nonstructural protein, NS_M) by the M RNA; and the N protein by the S RNA, Using two very different strategies, the S RNA, in addition, codes for a nonstructural protein (NS_S) (BISHOP 1985)

Early electron-microscopic studies (MURPHY et al. 1968) showed that virus particles mature intracellularly by budding into smooth vesicles in a perinuclear region of infected mice brain cells and tissue culture cells (for references see BISHOP and SHOPE 1979; PETTERSSON et al. 1988). Hardly ever have bunyaviruses been seen to bud at the PM. However, a strain of the phlebovirus Rift Valley fever virus was found to mature both intracellularly and at the PM in primary rat hepatocytes (ANDERSON and SMITH 1987). This remains the only reported example of budding at a site other than the GC.

Uukuniemi virus, a member of the *Uukuvirus* genus, has been used as a model to study the maturation process (VON BONSDORFF et al. 1970; KUISMANEN et al. 1982, 1984, 1985; GAHMBERG et al. 1986a, b; PETTERSSON et al. 1988). Using markers for the GC, immunofluorescence and immunoelectron microscopy of Uukuniemi virus-infected cells have shown that virus maturation occurs in the GC (KUISMANEN et al. 1982, 1984, 1985). During infection, both G1 and G2; as well as N, probably in the form of nucleocapsids, accumulate in the GC (KUISMANEN et al. 1982, 1984). The helical nucleocapsids appear to line up underneath the membrane of distended Golgi vesicles. As G1 and G2 accumulate in the GC, progressively more nucleocapsid also seems to enter the Golgi region. Little if any N protein is seen associated with the ER or the PM. Thus, a specific interaction between nucleocapsids and membranes containing the viral glycoproteins seems to exist only in the GC. During Uukuniemi virus infection, the

GC typically undergoes a morphological change. The stack of flat cisternae is progressively distorted, and instead the GC vacuolizes and the vesicles become partly dispersed in the cytoplasm (KUISMANEN et al. 1984). Whether, in fact, there is an expansion of the Golgi membranes is not clear. The vacuolization is probably caused by the accumulation of the glycoproteins in the GC, as shown by using a temperature-sensitive mutant defective in virus maturation at the restrictive temperature (GAHMBERG et al. 1986a). Interestingly, it has been shown that the morphologically altered Golgi can still terminally glycosylate and transport to the PM the Semliki Forest virus glycoproteins (GAHMBERG et al. 1986b). Thus, an intact Golgi structure is not an absolute requirement for proper function. Budding of bunyaviruses is inhibited by the ionophore monensin (CASH 1982; KUISMANEN et al. 1985), whereas the association of the nucleocapsids with the Golgi vesicles seems to be unaffected (KUISMANEN et al. 1985). The fact that monensin, which exchanges protons for sodium ions, blocks virus budding, points to the possibility that the pH or the ionic milieu prevailing in the GC is critical for bunyavirus budding. The budding in the GC may thus not only be dependent on a certain concentration of glycoproteins, but also on a conformational change of the glycoproteins induced by the milieu present in the GC. It is not known if virus budding occurs within the whole GC or just in a subcompartment. Vesicles containing one to several virus particles are thought to be transported to the PM, where virions are released upon fusion of the vesicles with the PM. The nature of the exocytic transport vesicles and their relation to normal exocytic trafficking is unknown.

3.2.2 The G1, G2, and NS_M Glycoproteins

All bunyaviruses possess two major glycoproteins encoded by the M RNA segment. The protein migrating more slowly on an SDS gel is referred to as G1, whereas the smaller is called G2. In cells infected with members of the *Bunyavirus* and *Phlebovirus* genera, a nonstructural protein NS_M derived from the M RNA segment has also been identified. The existence of NS_M has been confirmed by direct sequencing (see below) and identification of the product in infected or recombinant vaccinia virus-infected cells using antibodies against synthetic peptides (WASMOEN et al. 1988; KAKACH et al. 1989). Uukuniemi virus does not have an NS_M (RONNHOLM and PETTERSSON 1987) and the same seems to be true also for hantaviruses (SCHMALJOHN et al. 1987).

The complete sequence of the M RNA segment has been determined for Bunyamwera (BUN; LEES et al. 1986), La Crosse (LAC; GRADY et al. 1987), snowshoe hare (SSH; ESHITA and BISHOP 1984; FAZAKERLY et al. 1988) and Germiston (GER; PARDIGON et al. 1988) (*Bunyavirus* genus); Punta Toro (PT; IHARA et al. 1985) and Rift Valley fever (RVF; COLLETT et al. 1985) (*Phlebovirus* genus); Hantaan (HTN; SCHMALJOHN et al. 1987; YOO and KANG 1987) (*Hantavirus* genus); and Uukuniemi (UUK; RÖNNHOLM and PETTERSSON 1987) (*Uukuvirus* genus). In all four genera, the glycoproteins are synthesized principally as a precursor from one single open reading frame (ORF). This precursor has not been detected in infected cells, but in vitro translation of UUK and RVF virus M mRNA has yielded a 110-kDa (p100; UUK virus) or a 130-kDa (RVF virus) product that when made in the presence of microsomal membranes was cleaved into G1 and G2 (ULMANEN et al. 1981; SUZICH and COLLETT 1988). Processing of the precursor is thus likely to be contranslational (KUISMANEN 1984). Based on the sequence information, it seems that both glycoproteins are preceded by a separate signal sequence and that cleavage at the N terminus is carried out by a signal peptidase. Only for G2 of SSH virus (FAZAKERLY et al. 1988) has the exact cleavage site at the C terminus been determined. Therefore, the presence between the glycoproteins of short internal intergenic peptides, similar to the 6-kDa peptide of alphaviruses, cannot be excluded.

Sequence comparisons of the large ORF reveal a high degree of amino acid homology among members of the *Bunyavirus* genus (e.g., 89% between SSH and LAC and 43% between SSH and BUN viruses). The positions of the cysteines are highly conserved and the hydropathy profiles are almost indistinguishable. No apparent homology has been detected between members of the *Bunyavirus* genus and members of other genera. Within the *Phlebovirus* genus, RVF and PT display clear homology (35% for the N-terminal protein and 49% for the C-terminal one). Interestingly, there is low but probably significant homology (about 20%) between UUK virus G1 and G2 and the corresponding proteins in RVF and PT. Notably, about 70% of the cysteines have identical positions in all three viruses, suggesting a conserved three-dimensional structure of G1 and G2.

3.2.2.1 The Puzzling NS_M

The N termini of mature G2 (located N terminal of G1) of RVF virus and G1 (N terminal of G2) of PT virus are preceded by five and 13 in-frame initiation codons respectively. This suggests that the mature proteins could be preceded by prepeptides of approximately 17 kDa and 30 kDa respectively. In vitro mutagenesis followed by expression of the mutants from a vaccinia virus vector and analyses of the products using peptide antibodies indicates that the first AUG in RVF virus is used to produce a 78-kDa product, whereas the second AUG is utilized for the synthesis of a 14-kDa product (WASMOEN et al. 1988; KAKACH et al. 1989). The 78-kDa product contains the whole of G2, the 14-kDa peptide, and some upstream amino acids. Interestingly, the 78-kDa protein does not appear to be the precursor for G2. Thus, G2 and the 14-kDa presequence may be generated by cotranslational proteolytic cleavage of the protein initiated at the second AUG. WASMOEN et al. (1988), using synthetic peptide antibodies, have located both the 78-kDa and the 14-kDa NS_M proteins to the GC and to a reticular network, which could represent the ER, in cells transfected with mutant cDNAs expressing the proteins. The first and the fourth (or fifth) methionine codon precede typical signal peptide-like sequences that could direct the 78-kDa and G2 proteins through the ER membrane. The second and third methionine codons are not, however, followed by such hydrophobic amino acid sequences, making it unclear how the 14-kDa protein could be translocated through the ER membrane. It is therefore unclear whether the 14-kDa protein is indeed a membrane protein. The finding that the potential *N*-glycosylation site at Asn_{88} is utilized only in the 78-kDa protein, not in the 14-kDa protein, suggests that the latter protein is not translocated through the ER membrane.

Much less is known about the generation of the 30-kDa NS_M of PT virus. This presequence shows no apparent sequence homology with the 14-kDa NS_M presequence of RVF virus and has not yet been detected in infected cells or in virions.

In the M RNA segment of SSH virus there is an intergenic region between the G2 and G1 glycoproteins comprising 174 amino acids (about 19 kDa) that could conceivably be cleaved from the 1441-residue precursor as an NS_M polypeptide. Using peptide antibodies, 11-kDa and 10-kDa products have been detected in infected cells, but their exact relationship to the 174-residue sequence is unclear (FAZAKERLY et al. 1988).

So far no function has been attributed to the NS_M proteins. One possibility is that their primary function is simply to provide a signal sequence for the downstream glycoprotein, similar to the E3 protein (in case of RVF and PT viruses) or the 6-kDa peptide (in the case of SSH virus) of alphaviruses (MELANCON and GAROFF 1986). In contrast to the alphaviruses, where E3 is cleaved from the p62 precursor at a late stage of intracellular transport, NS_M of RVF and PT viruses appears to be cotranslationally cleaved. A role for NS_M in the intracellular transport or proper assembly of the spikes cannot, however, be excluded. Deletion of NS_M from the M RNA of RVF and PT does not seem to affect the intracellular transport or the targeting to the GC of G1 and G2 (WASMOEN et al. 1988; MATSUOKA et al. 1988). Expression of G1 and G2 together from cDNA cloned into a vaccinia vector indicates that the signal for targeting to the GC residues in G1 and/or G2 (see below).

3.2.2.2 G1 and G2

Based on the primary sequences of G1 and G2, it has been deduced that both proteins are oriented in the lipid bilayer as typical class I membrane proteins, i.e., with the N termini sticking out from the membrane and the C termini facing the nucleocapsid and the cytoplasm. In both cases, the bulk of the protein is located on the outside of the membrane forming the spikes. Both proteins are anchored to the lipid bilayer by a single transmembrane domain (PETTERSSON et al. 1988). In the case of UUK virus, treatment of microsomal vesicles from infected cells has confirmed this topology (N. Gahmberg and R.F. Pettersson, unpublished data). The cytoplasmic tail of G1 is at least 70 residues long and may be up to 96 residues long if the internal signal sequence of G2 remains attached to G1. In contrast, G2 is only five residues long (RÖNNHOLM and PETTERSSON 1987). The long G1 tail makes this protein a logical candidate for the interaction with the nucleocapsids. In all viruses, both G1 and G2 are very rich in cysteines (5%–7%), and as mentioned above the positions of the cysteines are highly conserved within the same genus. The detailed structure of the glycans has been analyzed only for UUK and Inkoo viruses (PESONEN et al. 1982a, b). Three types of structures were found in both viruses: a terminally glycosylated (sialylated) complex type, a typical high-mannose form (endo H-sensitive), and an intermediate type (endo H-resistant) which possibly represents an intermediate in the maturation from high-mannose to complex glycans. In UUK virus, G1 contains only endo H-resistant glycans, whereas G2 contains mainly endo H-sensitive ones. In HTN virus, the glycans are mostly of the high-mannose type (SCHMALJOHN et al. 1987). The presence of the immature glycans may reflect the site of maturation of bunyaviruses in the GC. The glycopeptides of California encephalitis viruses have also been characterized (VORNDAM and TRENT 1979).

The intracellular transport of G1 and G2 has been studied in some detail for UUK virus. The transport from the ER to the medial-Golgi has been analyzed by pulse-chase experiments using acquisition of endo H resistance as a marker. Half of G1 acquired endo H-resistant glycans in about 45 min (KUISMANEN 1984; GAHMBERG et al. 1986a) three to four times slower than, for example, the G protein of VSV or the HA of influenza virus. Studies with LAC virus have also indicated slow acquisition of endo H-resistant glycans (MADOFF and LENARD 1982). Recent results obtained with UUK virus indicate that G1 and G2 are transported from the ER to the site of virus budding at different speeds. Newly synthesized G1 is incorporated into virions some 20–30 min faster than G2 (KUISMANEN 1984; PERSSON and PETTERSSON, unpublished data). This also leads to an uneven distribution of G1 and G2 in the cell. In mid-infection of some 80% of G1, but only about 50% of G2, is localized to the GC. In virions, G1 and G2 are present in equimolar amounts as heterodimers (PERSSON et al. 1989). Since G1 and G2 are synthesized in equimolar amounts, these findings may mean that G1 and G2 are initially transported independently of each other. When and where dimerization occurs is not yet known.

Several lines of evidence suggest that G1 and G2 are retained in the GC. During infection large amounts of G1 and G2 accumulate in the GC and these proteins cannot be chased out of this organelle even during a 6-h-treatment with cycloheximide (GAHMBERG et al. 1986a).

Quantitation has indicated that less than 5% of the total content of G1 and G2 in UUK virus-infected cells is expressed on the cell surface (E. Kuismanen, unpublished data). However, in UUK virus-infected cells some glycoproteins are progressively expressed on the cell surface late in infection. The presence of cell-bound extracellular virus makes accurate quantitations difficult. Some glycoproteins have also been detected by immunoelectron microscopy on the surface of cells infected with a strain of RVF virus that exceptionally also buds at the PM (ANDERSON and SMITH 1987). Only trace amounts of G1 and G2 of PT, RVF, and HTN viruses expressed from cloned cDNA with the aid of vaccinia vectors were detected on the cell surface (MATSUOKA et al. 1988; WASMOEN et al. 1988; PENSIERO et al. 1988). Most of the glycoproteins accumulated in the GC and could not be chased out from there. When cells were double-infected with the UUK virus temperature-sensitive ts12 mutant (defective in virus maturation at 39°) and wild-type Semliki Forest virus (SFV), it was found that the SFV glycoproteins were readily transported to the PM, where budding of SFV occurred. In the same cells, the UUK virus glycoproteins were efficiently retained in the GC (GAHMBERG et al. 1986b). Taken together, all these results clearly show that the information necessary for Golgi retention is contained in the G1 and/or G2 proteins and that additional viral proteins are not needed. So far no information is available on the nature of this retention signal. There are several possibilities. The retention signal could be present in either G1 or G2, or both proteins could have an independent retention signal. In the former case, the protein lacking the signal would become Golgi-associated if it bound to (dimerized) the other protein. Alternatively, one could speculate that no clear-cut retention signal in the form of a linear amino acid sequence or a conformational domain exists, but that the proteins would become confined to the GC because of lateral protein–protein interactions forming large aggregates (arrays) in the GC which would be inhibited from further transport. The fact that the spikes on, for example, UUK virus particles are organized in an icosahedral lattice (VON BONSDORFF and PETTERSSON 1975) indicates that spike proteins may interact with each other. It should be noted that bunyaviruses lack a matrix (M) protein as found underneath the lipid bilayer of, for instance, rhabdoviruses, paramyxoviruses, and orthomyxoviruses. Expression of G1 and G2 individually will hopefully throw some light on these problems.

3.3 Poxviruses

Poxviruses are the largest (diameter about 200–400 nm) and most complex viruses known. Since the early 1960s it has been known that poxviruses develop in the cytoplasm and acquire an envelop in the GC. The best-studied representative is vaccinia virus, which belongs to the Orthopoxvirus genus of the Poxviridae family, and only this virus will be discussed here. For details on the replication and morphogenesis of poxviruses, the reviews by MORGAN (1976), DALES and POGO (1981), MOSS (1985), and the recent monograph on orthopoxviruses by FENNER et al. (1989) should be consulted.

Vaccinia virus infection results in the production of two types of infectious particles. Large numbers of "intracellular naked vaccinia" (INV) particles accumulate in the cytoplasm of infected cells. Most of the electron microscopy carried out during the past 25 years has focussed on the morphogenesis of INV particles. Only during the past 10 years has more attention been paid to the other form, the "extracellular enveloped vaccinia" (EEV) particles, which acquire an envelope in the GC. The INV particles are formed in a complex process steered by as yet unknown mechanisms. The morphogenesis of these particles has been excellently described in the recent review by FENNER et al. (1989). In summary, the morphogenically discernible steps are as follows. INV particle formation takes place within circumscribed, granular, electron-dense regions of the cytoplasm, called "viral factories" or B-type inclusion bodies. First, 50 to 55-nm-wide membranes (called caps or cupules) are formed in the factories. The cupules are completed to form roughly spherical "immature particles" containing a granular viroplasm. The membrane of the immature particles becomes the outer membrane of the virion, and the internal structures—the lateral bodies and the core—develop within this membrane. At an early stage, the membrane is covered by a layer of spicules consisting of a 65-kDa protein (ESSANI et al. 1982). The formation of the spicules is inhibited by rifampicin resulting in particles of irregular shape and size. Upon removal of rifampicin, spicules appear within minutes and the membrane assumes a regular convex shape (GRIMLEY et al. 1970). During further maturation, the spicules are replaced by surface tubules (ESSANI et al. 1982). The INV particles contain about 6% lipids, which make up the outer membrane. This membrane is thought to be a virus-specified, de novo-synthesized membrane that therefore is unrelated to any of the cellular membranes and is not acquired by a budding process. Mature INV particles, consisting of more than 100 different proteins, finally move out from the factories and accumulate in adjacent areas, and in some poxvirus-infected cells form so-called A-type inclusion bodies. The INV particles are not released from the cells unless they become enveloped.

Depending on the virus strain and the cell type used, between 1% and 30% of the INV particles go on to become enveloped in the GC (ICHIHASHI et al. 1971; MORGAN 1976; PAYNE 1979). The process of envelopment of EEV is clearly different from normal budding. According to electron-microscopic studies, the INV particle attaches to the cytoplasmic face of a Golgi cisternal membrane, apparently modified by viral proteins, and wraps a cisterna around itself. This leads to the acquisition of a double-membrane envelope. These virions migrate to the PM, where the outer membrane (equivalent to the membrane of an exocytic transport vesicle) fuses with the PM, thereby releasing the EEV particle from the cell. The EEV particle is thus composed of the INV surrounded by one Golgi-derived unit membrane. The formation of the double-membrane particles is blocked by monensin. Cytochalasin D, on the other hand, does not inhibit this step, but instead blocks the release of EEVs from the cell (PAYNE and KRISTENSSON 1982).

The EEV particles contain a 41-kDa (formerly described as 37-kDa) nonglycosylated protein and at least eight glycoproteins (PAYNE 1978, 1979). The 41-kDa protein, which is the major component of the envelope, has been localized to the Golgi complex and is covalently linked to palmitic acid (HILLER and WEBER 1985). Its sequence has recently been determined from the cloned gene (HIRT et al. 1986). The glycoproteins have molecular weights of 210 kDa, 110 kDa, 89 kDa, 41 kDa, and four between 23 and 28 kDa (formerly stated as 23–30 kDa). All these glycoproteins are also present on the cellular membranes, including the PM, of infected cells (PAYNE 1979). The 89-kDa protein is the viral HA (PAYNE 1979) and is also present on the cell surface, where it can be detected by the hemadsorption test. Its gene has been mapped and sequenced (SHIDA 1986). The EEV envelope is important for vaccinia pathogenesis. Antibodies to INV are not protective in vitro or in vivo, whereas antibodies specific for the envelope proteins protect against virus spread both in vitro and in vivo (PAYNE 1980).

The molecular dissection of vaccinia virus assembly is clearly lagging behind the descriptive morphogenic studies. This is not surprising in view of the complexity of the virus structure and the assembly process itself. Thus, no information is as yet available to explain what factors determine the site of envelopment of poxviruses in the GC, although the membrane-bound 41-kDa protein located in the Golgi membranes may play an important role in this process.

4 Virus Assembly at the Inner Nuclear Membrane

4.1 Herpesviruses

Early electron-microscopic studies showed that the herpesviruses, which include herpes simplex virus (HSV) types 1 and 2, Epstein-Barr virus (EBV), varicella-zoster virus (VZV), and cytomegalovirus (CMV), mature at the inner nuclear membrane. The first studies were done with HSV as early as the 1950s and early 1960s (MORGAN et al. 1954, 1959; DARLINGTON and MOSS 1968). Because HSV grows to high titers in tissue cultures, also later studies have utilized this virus as a model for studying herpesvirus envelopment. Since information on the assembly of other members of the Herpesviridae family is very scant, only HSV assembly is discussed below. The herpesviruses have been recently reviewed by ROIZMAN and BATTERSON (1985), and SPEAR (1980, 1984). The assembly process has also recently been discussed by STEPHENS and COMPANS (1988).

The assembly of herpesviruses is complex and as yet not understood in detail. The large genomic double-stranded DNA replicates in the nucleus, and the internal nucleocapsid proteins must be imported into the nucleus so that assembly of the nucleocapsid can take place in the nucleoplasm. Empty viral capsids are first assembled and accumulate in the nucleus. The DNA is then packaged into the empty capsids and simultaneously processed to unit length genomes. The nucleocapsids associate with sites at the inner nuclear membrane that have been modified by viral glycoproteins. Such sites of viral budding are characterized by a thickening of the membrane and of spikes, visible by EM, protruding into the perinuclear space. In some cases budding of HSV has been reported to take place at cytoplasmic membranes, but the significance of these observations remain unclear (for references see SPEAR 1980; ROIZMAN and BATTERSON 1985).

The envelope of mature extracellular HSV contains a number of glycoproteins. To date genetic loci for at least six different HSV-1 and five HSV-2 membrane glycoproteins have been identified (SPEAR 1984). It is likely that a few more will be found with time. The proteins are named gB1, gC1, gD1, gE1, gG1, gH1 (HSV-1) and gB2, gC2, gD2, gE2, gG2 (HSV-2). The precise functions of the individual proteins are still poorly known. gB is needed for infectivity and is involved in viral entry and fusion of the viral envelope with the host cell membranes (SARMIENTO et al. 1979; LITTLE et al. 1981). gC is the receptor for the C3b complement component and is not required for viral maturation or infectivity (FRIEDMAN et al. 1984). gE is the receptor for the Fc portion of immunoglobulins (BAUCKE and SPEAR 1979), and gD is the major target for neutralizing antibodies (LASKY et al. 1984; PARA et al. 1985) and may also be involved in cell fusion. gG also induces protection against infection (SULLIVAN and SMITH 1987). In fact, it seems that antibodies to all membrane glycoproteins can exhibit neutralizing activity.

Tunicamycin (TM), which inhibits the transfer of glycans to asparagine residues (*N*-linked glycans) blocks the production of infectious virus and results in the rapid degradation of the glycoproteins (PIZER et al. 1980). In the presence of TM, only gC acquires *O*-linked glycans, although all glycoproteins appear to contain this class of glycans (WENSKE and COURTNEY 1983). Apparently gC can be transported to the site of *O*-glycan chain addition, which is thought to occur in the trans-Golgi region (see Sect. 3.1.2), whereas the exit from the ER of the other glycoproteins is inhibited in the presence of TM. In the presence of TM, nucleocapsids accumulate in the nucleus, but also in the cytoplasm. The small number of particles formed under these conditions contained no detectable glycoproteins (PEAKE et al. 1982). Monensin, which blocks protein transport through the GC, inhibits the processing of HSV-1 glycoproteins and their transport to the cell surface (JOHNSON and SPEAR 1982). Addition of *O*-linked oligosaccharides did not occur in monensin-treated cells (JOHNSON and SPEAR 1983), suggesting that *O*-glycosylation occurs at a site distal to the action of monensin, which is thought to be at the level of medial-Golgi (GRIFFITHS et al. 1983). Production of extracellular virus was also efficiently reduced by monensin. Instead, infectious virus accumulated intracellularly in large vacuoles presumably derived from the GC. Such particles contained immature *N*-linked glycans. Such results were interpreted to mean that the egress of virus particles to the cell surface occurs via the GC (JOHNSON and SPEAR 1982).

Efforts have been made to localize the glycoproteins in herpesvirus-infected cells and to study their intracellular transport. Since budding occurs at the inner nuclear membrane, one must assume that all virion glycoproteins are transported from the site of synthesis in the ER to this membrane. This means that the glycoproteins must move from the outer nuclear membrane, which is contiguous with the ER, to the inner nuclear membrane. A likely, although not proven, mechanism is that the glycoproteins enter the nucleus via the nuclear pores.

Immunofluorescence, immunoelectron microscopy, and subcellular fractionation studies have indicated that the HSV glycoproteins are located in the nuclear membrane, although in most cases it has not been possible to distinguish whether they are present in the outer or the inner membrane (SPEAR 1980; COMPTON and COURTNEY 1984; STEPHENS and COMPANS 1988). The major EBV glycoprotein gp 350/220 was recently also localized by immunolabeling of freeze-fractured membranes to the inner nuclear membrane in addition to other cellular membranes (TORRISI et al. 1989). A systematic analysis of the localization of the glycoproteins by high-resolution immunoelectron microscopy is still lacking. In addition to the nuclear membrane, the glycoproteins have been found in the ER, GC, and PM. Thus, although primarily targeted to the inner nuclear membrane, all HSV glycoproteins containing the fully processed complex type of glycans are eventually found on the PM. Whether such proteins are transported to the PM directly via the GC or via the inner nuclear membrane is not known. HSV glycoproteins contain both *N*-linked and *O*-linked glycans (SPEAR 1984), and at least gE is in addition fatty acylated (JOHNSON and SPEAR 1983). The *N*-linked glycans are acquired in the ER, whereas the *O*-linked ones have been claimed to be added in the late Golgi region (JOHNSON and SPEAR 1983) (see also Sect. 3.1.2 for coronavirus E1). Most of the glycoproteins associated with the nuclear membrane contain high-mannose type of glycans. The high-mannose forms are only slowly processed to complex chains, and the transport of the glycoproteins to the PM is also unusually slow, occurring only after hours during a pulse-chase (cited in JOHNSON and SMILEY 1985). Interestingly, a marked difference in the kinetics of transport of gC expressed from a cloned gene was observed between uninfected and HSV-infected cells (JOHNSON and SMILEY 1985). In uninfected cells the processing of the glycans of gC was rapid (T1/2 about 20 min), whereas in infected cells it was delayed (T1/2 > 60 min). The appearance of gC on the cell surface was likewise rapid in uninfected cells: most of the protein was present on the cell surface 20 min after synthesis. When expressed in infected cells, gC appeared on the cell surface only after about 90–120 min. This could mean that gC is targeted to the nuclear membrane only when complexed to other viral glycoproteins. Expressed alone, gC would not be targeted to the nucleus, but instead would be rapidly transported to the PM via the GC.

Much attention has been focused on the gB protein, which has been suggested to be important in modifying intracellular membranes in infected cells. gB of HSV is homologous to the p110 glycoprotein of EBV (GONG et al. 1987). It has a 29-residue cleavable N-terminal signal sequence, followed by 696 residues of an external (luminal) domain and a 69-residue hydrophobic region that is thought to span the lipid bilayer three times, similarly to the coronavirus E1 (see Sect. 3.1.2). A 109-residue-long hydrophilic cytoplasmic tail is located at the C terminus (PELLETT et al. 1985; STUVE et al. 1987). In this domain, the sequence-Arg-Lys-Arg-Arg-(-Arg-Arg-Arg-Arg-in p110 of EBV) is found that might serve as a nuclear targeting signal and nuclear matrix binding sequence similar to the ones found in the large T antigens of several nuclear proteins (DINGWALL 1985; ADAM et al. 1989), notably the large T antigen of SV40 (KALDERON et al. 1984) and polyoma (RICHARDSON et al. 1986). The possible role of this basic sequence in translocation to the nucleus has not yet been tested by in vitro mutagenesis. Within minutes after synthesis, gB dimerizes and may also form higher-order oligomers (CLAESSON-WELSH and SPEAR 1986, 1987). It contains both *N*-linked and *O*-linked glycans. In infected cells and when expressed from the cloned gene, gB localizes to the nuclear membrane, ER, GC, and PM (ALI et al. 1987; PACHL et al. 1987). In contrast to gB, p110 of EBV has not been found on the surface of infected cells. Most of p110 was localized by immunoelectron microscopy to the outer and inner nuclear membranes and to the ER (GONG et al. 1987). Like gB, gG of HSV-1 expressed from a recombinant vaccinia virus was also localized to the nuclear membrane in addition to the PM (SULLIVAN and SMITH 1987).

What, then determines the site of herpesvirus budding? At present there is no clear answer to this question. It is obvious that the virion glycoproteins have to be targeted to the inner nuclear membrane. It is possible that only one, or a few, of the many virus-specified glycoproteins contains a nuclear targeting signal. In that case, the other glycoproteins lacking a signal would have to complex with the signal-containing glycoprotein(s) in order to reach the inner nuclear membrane. Alternatively, a targeting signal might be created only after a complex between the viral glycoproteins has been assembled. A third possibility is that all glycoproteins have their own nuclear targeting signal. The role of the nucleocapsid and viral tegument proteins in the assembly process has yet to be explored. One possibility is that one or more of the core proteins containing a nuclear translocation signal binds to cytoplasmic domains of the glycoproteins and guides their transport into the nucleus; however, no evidence to support this hypothesis has yet been produced. Thus, in conclusion, at present little information is available to explain the mechanisms by which virus budding occurs at this unique site. Once budding has occurred, virus particles are present initially in the perinuclear space between the inner and outer nuclear membrane, i.e., in the lumen of the ER. From there, particles are evidently transported to the GC and further to the PM, where they are released upon fusion of transport vesicles with the PM. *N*-linked and *O*-linked glycans appear to be processed on the virus particles during this transport (SPEAR 1984).

5 Concluding Remarks

In this review, examples of enveloped viruses that mature at intracellular membranes by a budding process have been discussed. The membranes utilized by such viruses include the endoplasmic reticulum (ER), the Golgi complex (GC), and the inner nuclear membrane. The central question is what factors determine the site of virus budding. In the case of viruses that bud at the plasma membrane (PM), the viral glycoproteins are rapidly transported to the PM via the normal exocytic pathway (STEPHENS and COMPANS 1988). In all respects, these proteins behave like any host cellular PM protein. In contrast, for most of the intracellularly maturing enveloped viruses at least one of the virus-specified glycoproteins is targeted to and accumulates in the budding compartment. Examples of such glycoproteins are NS28 and VP7 of rotaviruses (ER), E1 of coronaviruses (an intermediate compartment between ER and GC, or GC), and G1 and G2 of bunyaviruses (GC). For herpesviruses, the glycoproteins are targeted to the inner nuclear membrane but are also transported to the PM. Thus, one important factor in determining the site of budding is clearly the targeting to and accumulation of glycoproteins in the budding compartment. The question then arises of what directs the proteins to a particular compartment and how the proteins are retained there. There are several possibilities: (a) The proteins may contain stretches of amino acids (linear or conformational) that constitute a "retention signal". Such a mechanism seems to operate in the case of the VP7 protein of rotaviruses (which, however, is not a transmembrane protein) and E1 of coronaviruses. Recently, a retention signal for the ER membrane has also been defined at the C terminus of the nonstructural adenovirus-encoded membrane protein E3/19k (Pääbo et al. 1987; R. Gabathuler, B. Dahllöf and S. Kvist, personal communication). Similar retention sequences have not yet been identified for any other viral or cellular transmembrane proteins. It is still not known what cellular structures, if any, these retention sequences interact with. (b) Lateral interaction between viral membrane proteins, resulting in the formation of large aggregates, may exclude the proteins from transport vesicles and thus cause them to accumulate in a particular compartment along the exocytic pathway. (c) Conformational changes of the glycoproteins or the acquisition of the right oligomeric structures may occur along the transport route. Such changes may take place because of changes in the milieu (pH, ionic conditions) or because of posttranslational modifications (e.g., processing of glycans). These changes may be necessary before an interaction with the nucleocapsids can take place. (d) An important prerequisite for virus budding may be the need for a critical concentration of viral glycoproteins. A slow intracellular transport could result in the accumulation of viral glycoproteins in the ER or the GC above a critical level needed for budding. (e) The nucleocapsids may play an important role in determining the budding site. For instance, the nucleocapsids of bunyaviruses must gain access to and be transported into the Golgi region, from which ribosomes, for example, are normally excluded. It is possible that active transport of the nucleocapsids to this region occurs. On the other hand, nucleocapsids may be competent to interact with glycoproteins having a particular conformation or structure, which they acquire only in a particular compartment [see also (c) above]. (f) Finally, in the case of flaviviruses, the assembly of the nucleocapsids may occur in association with the ER membrane. In this case, the nucleocapsids would determine the budding site. This process may be closely linked to both protein and RNA synthesis. It is reasonable to assume that the viruses discussed here have evolved different mechanisms for budding at intracellular membranes.

One tempting speculation is that viral spike proteins have evolved from cellular proteins. If this is the case, then the glycoproteins discussed in this review could serve as useful models for studying the intracellular transport and targeting of compartment-specific cellular proteins (Pettersson et al. 1988). A final comment regards the question of whether intracellular maturation offers any particular advantage to the virus. This is difficult to envisage. Some years ago it was speculated that cells infected with intracellularly maturing viruses could somehow escape killing by cytotoxic T lymphocytes. In the light of recent progress in elucidating the mechanism by which cytotoxic T lymphocytes recognize processed viral antigens in the context of class I MHC antigens and kill infected cells, however, this hypothesis can be abandoned.

Acknowledgements. I am indebted to John Armstrong, Paul Atkinson, Richard Bellamy, Mark Collett, Richard Compans, Mary Estes, Franz Heinz, Heiner Niemann, Lendon Payne, Jack Rose, Peter Rottier, Connie Schmaljohn, Jonathan Smith, Willy Spaan, Patricia Spear, John and Sharon Tooze, and Gerd Wengler for communicating both published and unpublished results, and in some cases for critical reading of the manuscript. I also thank Ms Viveca Karlsson for typing the manuscript.

References

Adam SA, Lobl TJ, Mithell MA, Gerace L (1989) Identification of specific binding proteins for a nuclear location sequence. Nature 337: 276–279
Ali MA, Butcher M, Ghosh HP (1987) Expression and nuclear envelope localization of biologically active fusion glycoprotein gB of herpes simplex virus in mammalian cells using cloned DNA. Proc Natl Acad Sci USA 84: 5675–5679
Anderson GW Jr, Smith JF (1987) Immunoelectron microscopy of Rift Valley fever viral morphogenesis in primary rat hepatocytes. Virology 161: 91–100
Andrew ME, Boyle DB, Coupar BEH , Whitfeld PL, Both GW, Bellamy AR (1987) Vaccinia virus recombinants expressing the SAll rotavirus VP7 glycoprotein gene induce serotype-specific neutralizing antibodies. J Virol 61: 1054–1060
Armstrong J, Niemann H, Smeekens S, Rottier P, Warren G (1984) Sequence and topology of a model intracellular membrane protein, El glycoprotein, from a coronavirus. Nature 308: 751–752
Armstrong J, McCrae M, Colman A (1987) Expression of coronavirus E1 and rotavirus VP10 membrane proteins from synthetic RNA. J Cell Biochem 35: 129–136
Au K-S, Chan W-K, Estes MK (1989a) Rotavirus morphogenesis involves an endoplasmic reticulum transmembrane glycoprotein. In: UCLA symposium 88, cell biology of viral entry, replication and pathogenesis. Liss, New York, pp 257–267
Au K-S, Chan W-K, Burns JW, Estes MK (1989b) Receptor activity of the rotavirus nonstructural glycoproteins NS28. J Virol 63: 4553–4562
Baucke RB, Spear PG (1979) Membrane proteins specified by herpes simplex viruses. V. Identification of an Fc-binding glycoprotein. J Virol 32: 779–789
Baybutt HN, McCrae MA (1984) The molecular biology of rotavirus. VII. Detailed structural analysis of gene 10 of bovine rotavirus. Virus Res 1: 533–541
Bellamy AR, Both GW (1990) Molecular biology of rotaviruses. Adv Virus Res 38: 1–43
Bergmann CC, Maass D, Poruchynsky MS, Atkinson PH, Bellamy AR (1989) Topology of the non-structural rotavirus receptor glycoprotein NS28 in the rough endoplasmic reticulum. EMBO J 8: 1695–1703
Binns MM, Boursnell MEG, Cavanagh D, Pappin DJC, Brown TDK (1985) Cloning and sequencing of the gene encoding the spike protein of the coronavirus IBV. J Gen Virol 66: 719–726
Bishop DHL (1985) Replication of arenaviruses and bunyaviruses. In: Fields BN, Knipe DM, Chanock RM, Melnick JL, Roizman B, Shope RE (eds) Virology, Raven, New York, pp 1083–1110
Bishop DHL, Shope RE (1979) Bunyaviridae. Compr Virol 14: 1–156
Both GW (1988) Replication of reoviridae: information derived from gene cloning and expression. In: Domingo E, Holland JJ, Ahlquist P (eds) RNA genetics. CRC Press, Boca Raton, pp 172–193
Both GW, Mattick JS, Bellamy AR (1983a) Serotype-specific glycoprotein of simian 11 rotavirus: coding assignment and gene sequence. Proc Natl Acad Sci USA 80: 3091–3095
Both GW, Siegman LJ, Bellamy AR, and Atkinson PH (1983b) Coding assignment and nucleotide sequence of simian rotavirus SAll gene segment 10: location of glycosylation sites suggests that the signal peptide is not cleaved. J Virol 48: 335–339

Boulton RW, Westaway EG (1976) Replication of the flavivirus Kunjin: proteins, glycoproteins, and maturation associated with cell membranes Virology. 69: 416–430
Boursnell MEG, Brown TDK, Foulds IJ, Green PF, Tomley FM, Binns MM (1987) Completion of the sequence of the genome of the coronavirus avian infectious bronchitis virus. J Gen Virol 68: 57–77
Boyle JF, Weismiller DG, Holmes KV (1987) Genetic resistance to mouse hepatitis virus correlates with absence of virus-binding activity on target tissues. J Virol 61: 185–189
Brinton MA (1986) Replication of flaviviruses. In: Schlesinger S and Schlesinger MJ (eds) The Togaviridae and Flaviviridae. Plenum, New York, pp 327–374
Burns JW, Greenberg HB, Shaw RD, Estes MK (1988) Functional and topographical analyses of epitopes on the hemagglutinin (VP4) of the simian rotavirus Sa11. J Virol 62: 2164–2172
Cash P (1982) Inhibition of LaCrosse virus replication by monensin, a monovalent ionophore. J Gen Virol 59:193–196
Castle E, Nowak T, Leidner U, Wengler G, Wengler G (1985) Sequence analysis of the viral core protein and the membrane-associated proteins V1 and NV2 of the flavivirus West Nile virus and of the genome sequence for these proteins. Virology 145: 227–236
Castle E, Leidner U, Nowak T, Wengler G, Wengler G (1986) Primary structure of the West Nile flavivirus genome region coding for all nonstructural proteins. Virology 149: 10–26
Caul EO, Egglestone SI (1977) Further studies on human enteric coronaviruses. Arch Virol 54: 107–117
Cavanagh D (1983) Coronavirus IBV: structural characterization of the spike protein. J Gen Virol 64:2577–2583
Chan W-K, Penaranda ME, Crawford SE, Estes MK (1986) Two glycoproteins are produced from the rotavirus neutralization gene. Virology 151: 243–252
Chan W-K, Au K-S, Estes MK (1988) Topography of the Simian rotavirus nonstructural glyoprotein (NS28) in the endoplasmic reticulum membrane. Virology 164: 435–442
Chasey D (1980) Investigation of immunoperoxidase-labelled rotavirus in tissue culture by light and electron microscopy. J Gen Virol 50: 195–200
Chasey D, Alexander DJ (1976) Morphogenesis of avian infectious bronchitis virus in primary chick kidney cells. Arch Virol 52: 101–111
Claesson-Welsh L, Spear PG (1986) Oligomerization of herpes simplex virus glycoprotein B. J Virol 60: 803–806
Claesson-Welsh L, Spear PG (1987) Amino-terminal sequence, synthesis, and membrane insertion of glycoprotein B of herpes simplex virus type 1. J Virol 61: 1–7
Coia G, Parker MD, Speight G, Byrne ME, Westaway EG (1988) Nucleotide and complete amino acid sequence of Kunjin virus: definitive gene order and characteristics of the virus-specific proteins. J Gen Virol 69: 1–21
Collett MS, Purchio AF, Keegan K, Frazier S, Hays W, Anderson DK, Parker MD, Schmaljohn C, Schmidt J, Dalrymple J (1985) Complete nucleotide sequence of the M RNA segment of Rift Valley fever virus. Virology 144: 228-245
Compton T, Courtney RJ (1984) Virus-specific glycoproteins associated with the nuclear fraction of herpes simplex virus type 1-infected cells. J Virol 49: 594–597
Copeland CS, Doms RW, Bolzau EM, Webster RG, Helenius A (1986) Assembly of influenza hemagglutinin trimers and its role in intracellular transport. J Cell Biol 103: 1179–1191
Copeland CS, Zimmer K-P, Wagner KR, Healey Ga, Mellman I, Helenius A (1988) Folding, trimerization, and transport are sequential events in the biogenesis of influenza virus hemagglutinin. Cell 53: 197–209
Dales S, Pogo BGT (1981) Biology of poxviruses. In: Gard S et al. (eds) Virology monographs, vol 18. Springer, Vienna, New York, PP 1–109
Darlington RW, Moss LH (1968) Herpesvirus envelopment. J Virol 2: 48–55
Deregt D, Sabara M, Babiuk LA (1987) Structural proteins of bovine coronavirus and their intracellular processing. J Gen Virol 68: 2863–2877
Dingwall C (1985) The accumulation of proteins in the nucleus. Trends Biochem Sci 10: 64–66
Doms RW, Keller DS, Helenius A, Balch WE (1987) Role of adenosine triphosphate in regulating the assembly and transport of vesicular stomatitis virus G protein trimers. J Cell Biol 105: 1957–1969
Doughri AM, Storz J, Hajer I, Fernando HS (1976) Morphology and morphogenesis of a coronavirus infecting intestinal epithelial cells of newborn calves. Exp Mol Pathol 25: 355–370
Dubois-Dalcq M, Holmes RV, Rentier B (1984) Assembly of enveloped RNA viruses. Springer, Vienna, New York

Einfeld D, Hunter E (1988) Oligomeric structure of a prototype retrovirus glycoprotein. Proc Natl Acad Sci USA 85: 8688–8692

Ericson BL, Graham DY, Mason BB, Estes MK (1982) Identification, synthesis and modifications of simian rotavirus SA11 polypeptides in infected cells. J Virol 42: 825–839

Ericson BL, Graham DY, Mason BB, Hanssen HH, Estes MK (1983) Two types of glycoprotein precursors are produced by the simian rotavirus SA11. Virology 127: 320–332

Eshita Y, Bishop DHL (1984) The complete sequence of the M RNA of snowshoe hare bunyavirus reveals the presence of internal hydrophobic domains in the viral glycoprotein. Virology 137: 227–240

Espejo RT, López S, Arias C (1981) Structural polypeptides of simian rotavirus SA11 and the effect of trypsin. J Virol 37: 156–160

Essani K, Dugre R, Dales S (1982) Biogenesis of vaccinia: involvement of spicules of the envelope during virion assembly examined by means of conditional lethal mutants and serology. Virology 118: 279–292

Estes MK, Cohen J (1989) Rotavirus gene structure and function. Microbiol Rev 53: 410–449

Estes MK, Graham DY, Mason BB (1981) Proteolytic enhancement of rotavirus infectivity: molecular mechanisms. J Virol 39: 879–888

Estes MK, Palmer EL, Obijeski JF (1983) Rotaviruses: a review. In: Cooper M et al. (eds) Current topics in microbiology and immunology, vol 105. Springer, Berlin Heidelberg New York, pp 123–184

Estes MK, Crawford SE, Penaranda ME, Petrie BL, Burns JW, Chan W-K, Ericson B, Smith GE, Summers MD (1987) Synthesis and immunogenicity of the rotavirus major capsid antigen using a baculovirus expression system. J Virol 61: 1488–1494

Fazakerley JK, Gonzalez-Scarano F, Strickler J, Dietzschold B, Karush F, Nathanson N (1988) Organization of the middle RNA segment of snowshoe hare Bunyavirus. Virology 167: 422–432

Fenner F, Wittek R, Dumbell KR (1989) The orthopoxviruses. Academic, New York pp 1–432

Friedman HM, Cohen GH, Eisenberg RJ, Seidel CA, Cines DB (1984) Glycoprotein C of herpes simplex virus 1 acts as a receptor for the C3b complement component on infected cells. Nature 309: 633–635

Gahmberg N, Kuismanen E, Keränen S, Pettersson RF (1986a) Uukuniemi virus glycoproteins accumulate in and cause morphological changes of the Golgi complex in the absence of virus maturation. J Virol 57: 899–906

Gahmberg N, Pettersson RF, Kääriäinen L (1986b) Efficient transport of Semliki Forest virus glycoproteins through a Golgi complex morphologically altered by Uukuniemi virus glycoproteins. EMBO J 5: 3111–3118

Garoff H, Simons K (1974) Location of the spike glycoproteins in the Semliki Forest virus membrane. Proc Natl Acad Sci USA 71: 3988–3992

Geetha-Habib M, Noiva R, Kaplan HA, Lannarz WJ (1988) Glycosylation site binding protein, a component of oligosaccharyl transferase, is highly similar to three other 57 kd luminal proteins of the ER. Cell 54: 1053–1060

Glass RI, Keith J, Nakagomi O, Nakagomi T, Askaa J, Kapikian AZ, Chanock RM, Flores J (1985) Nucleotide sequence of the structural glycoprotein VP7 gene of Nebraska calf diarrhea virus rotavirus: comparison with homologous genes from four stains for human and animal rotaviruses. Virology 141: 292–298

Gong M, Ooka T, Matsuo T, Kieff E (1987) Epstein-Barr virus glycoprotein homologous to herpes simplex virus gB. J Virol 61: 499–508

Gonzalez-Scarano F, Nathanson N (1990) Bunyaviruses. In: Fields BN, Knipe DM, Chanock RM, Melnick JL, Roizman B, Shope RE (eds) Virology. Raven, New York, pp 1195–1226

Gorziglia M, Larrea C, Liprandi F, Esparza J (1985) Biochemical evidence for the oligomeric (possibly trimeric) structure of the major inner capsid polypeptide (45k) of rotaviruses. J Gen Virol 66: 1889–1900

Gorziglia M, Aguirre Y, Hoshino Y, Esparza J, Blumentals I, Askaa J, Thompson M, Glass RI, Kapikian AZ, Chanock RM (1986) VP7 serotype-specific glycoprotein of OSU porcine rotavirus: coding assignment and gene sequence. J Gen Virol 67: 2445–2454

Grady LJ, Sanders ML, Campbell WP (1987) The sequence of the M RNA of an isolate of La Crosse virus. J Gen Virol 68: 3057–3071

Greenberg HB, Valdesuso J, van Wyke K, Midthun K, Walsh M, McAuliffe V, Wyatt Rg, Kalica AR, Flores J, Hoshino Y (1983) Production and preliminary characterization of monoclonal antibodies directed at two surface proteins of rhesus rotavirus. J Virol 47: 267–275

Griffiths G, Quinn P, Warren G (1983) Dissection of the Golgi complex. I. Monensin inhibits the transport of viral membrane proteins from medial to trans Golgi cisternae in baby hamster kidney cells infected with Semliki Forest virus. J Cell Biol 96: 835–850

Grimley PM, Rosenblum EN, Mims SJ, Moss B (1970) Interruption by rifampicin of an early stage in vaccinia virus morphogenesis: accumulation of membranes which are precursors of virus envelopes. J Virol 6: 519–533

Gunn PR, Sato F, Powell KFH, Bellamy AR, Napier JR, Harding DRK, Hancock WS, Seigman LJ, Both GW (1985) Rotavirus neutralizing protein VP7: antigenic determinants investigated by sequence analysis and peptide synthesis. J Virol 54: 791–797

Hahn YS, Galler R, Hunkapiller T, Dalrymple JM, Strauss JH, Strauss EG (1988) Nucleotide sequence of dengue 2 RNA and comparison of the encoded proteins with those of other flaviviruses. Virology 162: 167–180

Helenius A, Kartenbeck J (1980) The effects of octylglucoside on the Semliki Forest virus membrane: evidence for a spike-protein-nucleocapsid interaction. Eur J Biochem 106: 613–618

Hiller G. Weber K (1985) Golgi-derived membranes that contain an acylated viral polypeptide are used for vaccinia virus envelopment. J Virol 55: 651–659

Hirt P, Hiller G, Wittek R (1986) Localization and fine structure of a vaccinia virus gene encoding on envelope antigen. J Virol 58: 757-764

Holmes IH (1983) Rotaviruses. In: Joklik WK (ed) The Reoviridae. Plenum, New York, pp 359–422

Holmes KV (1985) Replication of coronaviruses. In: Fields BN, Knipe DM, Chanock RM, Melnick JL, Roizman B, Shope RE (eds) Virology. Raven, New York, pp 1331–1343

Holmes KV, Doller EW, Sturman LS (1981) Tunicamycin resistant glycosylation of a coronavirus glycoprotein: demonstration of a novel type of viral glycoprotein. Virology 115: 334–344

Hunter E (1988) Membrane insertion and transport of viral glycoproteins: a mutational analysis. In: Das RC, Robbins PW (eds) Protein transfer and organelle biogenesis, Academic, New York, pp 109–158

Ichihashi Y, Matsumoto S, Dales S (1971) Biogeneis of poxviruses: role of A-type inclusions and host cell membranes in virus dissemination. Virology 46: 507–532

Ihara T, Smith J, Dalrymple JM, Bishop DHL (1985) Complete sequences of the glycoproteins and M RNA of Punta Toro phlebovirus compared to those of Rift Valley fever virus. Virology 144: 246–259

Ishak R, Tovey DG, Howard CR (1988) Morphogenesis of yellow fever virus 17D in infected cell cultures. J Gen Virol 69: 325–335

Johnson DC, Smiley JR (1985) Intracellular transport of herpes simplex virus gD occurs more rapidly in uninfected cells than in infected cells. J Virol 54: 682–689

Johnson DC, Spear PG (1982) Monensin inhibits the processing of herpes simplex virus glycoproteins, their transport to the cell surface, and the egress of virions from infected cells. J Virol 43: 1102–1112

Johnson DC, Spear PC (1983) O-linked oligosaccharides are acquired by herpes simplex virus glycoproteins in the Golgi apparatus. Cell 32: 987–997

Kabcenell AK, Atkinson PH (1985) Processing of the rough endoplasmic reticulum membrane glycoproteins of rotavirus SA11. J Cell Biol 101: 1270–1280

Kabcenell AK, Poruchynsky MS, Bellamy AR, Greenberg HB, Atkinson PH (1988) Two forms of VP7 are involved in assembly of SA11 rotavirus in endoplsmic reticulum. J Virol 62: 2929–2941

Kakach LT, Wasmoen TL, Collett MS (1988) Rift Valley fever virus M segment: use of recombinant vaccinia viruses to study Phlebovirus gene expression. J Virol 62: 826–833

Kakach LT, Suzich JA, Collett MS (1989) Rift Valley fever virus M segment: Phlebovirus expression strategy and protein glycosylation. Virology 170: 505–510

Kalderon D, Richardson WD, Markham AF, Smith AE (1984) Sequence requirements for nuclear location of simian virus 40 large-T antigen. Nature 311: 33–38

Kalica AR, Flores J, Greenberg HB (1983) Identification of the rotaviral gene that codes for hemagglutination and protease-enhanced plaque formation. Virology 125: 194–205

Kapikian AZ, Chanock RM (1985) Rotaviruses. In: Fields BN, Knipe DM, Chanock RM, Melnick JL, Roizman B, Shope RE (eds) Virology. Raven, New York, pp 863–905

Kelly RB (1985) Pathways of protein secretion in eukaryotes. Science 230: 25–32

Kozak M (1986) Bifunctional messenger RNAs in eukaryotes. Cell 47: 481–483

Kreis TE, Lodish HF (1986) Oligomerization is essential for transport of vesicular stomatitis viral glycoprotein to the cell surface. Cell 46: 929–937

Kuismanen E (1984) Posttranslational processing of Uukuniemi virus glycoproteins G1 and G2. J Virol 51: 806–812

Kuismanen E, Hedman K, Saraste J, Petterson RF (1982) Uukuniemi virus maturation: accumulation of virus particles and viral antigens in the Golgi complex. Mol Cell Biol 2: 1444–1458
Kuismanen E, Bång B, Hurme M, Pettersson RF (1984) Uukuniemi virus maturation: immunofluorescence microscopy with monoclonal glycoprotein-specific antibodies. J Virol 51: 137–146
Kuismanen E, Saraste J, Pettersson RF (1985) Effect of monensin on the assembly of Uukuniemi virus in the Golgi complex. J Virol 55: 813–822
Lasky LA, Dowbenko D, Simonsen CC, Berman PW (1984) Protection of mice from lethal herpes simplex virus infection by vaccination with a secreted form of cloned glycoprotein D. Biotechnology 2: 527–532
Lees JF, Pringle CR, Elliott RM (1986) Nucleotide sequence of the Bunyamwera virus M RNA segment: conservation of structural features in the bunyavirus glycoprotein gene product. Virology 148: 1–14
Little SP, Jofre JT, Courtney RJ, Schaffer PA (1981) A virion-associated glycoprotein essential for infectivity of herpes simplex virus type 1. Virology 115: 149–160
Liu M, Offit PA, Estes MK (1988) Identification of the simian rotavirus SA11 genome segment 3 product. Virology 163: 26–32
Luytjes W, Sturman LS, Bredenbeek PJ, Charite J, van der Zeijst BA, Horzinek MC, Spaan WJ (1987) Primary structure of the glycoprotein E2 of coronavirus MHV-A59 and identification of the trypsin cleavage site. Virology 161: 479–487
Machamer CE, Rose JK (1987) A specific transmembrane domain of a coronavirus E1 glycoprotein is required for its retention in the Golgi region. J Cell Biol 105: 1205–1214
Madoff DH, Lenard J (1982) A membrane glycoprotein that accumulates intracellularly: cellular processing of the large glycoprotein of La Crosse virus. Cell 28: 821–829
Massalski A, Coulter-Mackie M, Knobler RL, Buchmeier MJ, Dales S (1982) In vivo and in vitro models of demyelinating diseases. V. Comparison of the assembly of mouse hepatitis virus, strain JHM, in two murine cell lines. Intervirology 18: 135–146
Matsumura T, Shiraki K, Sashikata T, Hotta S (1977) Morphogenesis of Dengue-1 virus in cultures of a human leukemic leukocyte line (J-111). Microbiol Immuno1 21: 329–334
Matsuoka Y, Ihara T, Bishop DHL, Compans RW (1988) Intracellular accumulation of Punta Toro virus glycoproteins expressed from cloned cDNA. Virology 167: 251–260
Mayer T, Tamura T, Falk M, Niemann H (1988) Membrane integration and intracellular transport of the coronavirus glycoprotein E1, a class III membrane glycoprotein. J Biol Chem 263: 14956–14963
Melancon P, Garoff H (1986) Reinitiation of translocation in the Semliki Forest virus structural polyprotein: identification of the signal for the E1 glycoprotein. EMBO J 7: 1543–1550
Meyer JC, Bergmann CC, Bellamy AR (1989) Interaction of rotavirus cores with the nonstructural glycoprotein NS28. Virology 171: 98–107
Monath TP (1985) Flaviviruses. In: Fields BN, Knipe DM, Chanock RM, Melnick JL, Roizman B, Shope RE (eds) Virology. Raven, New York, pp 955–1004
Morgan C (1976) Vaccinia virus reexamined: development and release. Virology 73: 43–58
Morgan C, Ellison SA, Rose HM, More DH (1954) Structure and development of viruses as observed in electron microscope. I. Herpes simplex virus. J Exp Med 100: 195–202
Morgan C, Rose HM, Holden M, Jones EP (1959) Electron microscopic observations on the development of herpes simplex virus. J Exp Med 110: 643–656
Moss B (1985) Replication of poxviruses. In: Fields BN, Knipe DM,Chanock RM, Melnick JL, Roizman B, Shope RE (eds) Virology. Raven, New York, pp 685–703
Munro S, Pelham HRB (1987) A C-terminal signal prevents secretion of luminal ER proteins. Cell 48: 899–907
Murphy FA (1980) Togavirus morphology and morphogenesis. In: Schlesinger RW (ed) The togaviruses. Academic, New York, pp 241–316
Murphy FA, Whitfield SG, Coleman PH, Calisher CH, Rabin ER, Jenson AB, Melnick JL, Edwards MR, Whitney E (1968) California group arboviruses: electron microscopic studies. Exp Mol Pathol 9: 44–56
Murphy FA, Harrison AK, Whitfield SG (1973) Bunyaviridae: morphological and morphogenetic similarities of Bunyamwera serologic supergroup viruses and several other arthropod-borne viruses. Intervirology 1: 297–316
Niemann H, Klenk H-D (1981) Coronavirus glycoprotein E1, a new type of viral glycoprotien. J Mol Biol 153: 993–1010
Niemann H, Boschek B, Evans D, Rosing M, Tamura T, Klenk H-D (1982) Post-translational glycosylation of coronavirus glycoprotein E1: inhibition by monensin. EMBO J 1: 1499–1504

Niemann H, Geyer R, Klenk H-D, Linder D, Stirm S, Wirth M (1984) The carbohydrates of mouse hepatitis virus (MHV) A59: structures of the O-glycosidically linked oligosaccharides of glycoprotein E1. EMBO J 3: 665–670

Nowak T, Färber PM, Wengler G, Wengler G (1989) Analyses of the terminal sequences of West Nile virus structural proteins and of the in vitro translation of these proteins allow to propose a complete scheme of the proteolytic cleavages involved in their synthesis. Virology 169: 365–376

Offit PA, Blavat G, Greenberg HB, Clark HF (1986) Molecular basis of rotavirus virulence: role of gene segment 4. J Virol 57: 46–49

Okada Y, Richardson MA, Ikegami N, Nomoto A, Furuichi Y (1984) Nucleotide sequence of human rotavirus genome segment 10, an RNA encoding a glycosylated virus protein. J Virol 51: 856–859

Pääbo S, Bhat BM, Wold WS, Peterson PA (1987) A short sequence in the COOH-terminus makes an adenovirus membrane glycoprotein a resident of the endoplasmic reticulum. Cell 50: 311–317

Pachl C, Burke RL, Stuve LL, Sanchez-Pescador L, van Nest G, Masiarz F, Dina D (1987) Expression of cell-associated and secreted forms of herpes simplex virus type 1 glycoprotein gB in mammalian cells. J Virol 61: 315–325

Palade G (1975) Intracellular aspects of the process of protein secretion. Science 189: 347–358

Para MF, Parish ML, Noble AG, Spear PG (1985) Potent neutralizing acitivity associated with anti-glycoprotein D specificity among monoclonal antibodies selected for binding to herpes simplex virions. J Virol 55: 483–488

Pardigon N, Vialat P, Gerbaud S, Girard M, Bouloy M (1988) Nucleotide sequence of the M segment of Germiston virus: comparison of the M gene product of several bunyaviruses. Virus Res 11: 73–85

Payne LG (1978) Polypeptide composition of extracellular enveloped vaccinia virus. J Virol 27: 28–37

Payne LG (1979) Identification of the vaccinia hemagglutinin polypeptide from a cell system yielding large amounts of extracellular enveloped virus. J Virol 31: 147–155

Payne LG (1980) Significance of extracellular enveloped virus in the in vitro and in vivo dissemination of vaccinia. J Gen Virol 50: 89–100

Payne LG, Kristensson K (1982) Effect of glycosylation inhibitors on the release of enveloped vaccinia virus. J Virol 41: 367–375

Peake ML, Nystrom P, Pizer LI (1982) Herpesvirus glycoprotein synthesis and insertion into plasma membranes. J Virol 42: 678–690

Pelham HRB (1988) Evidence that luminal ER proteins are sorted from secreted proteins in a post-ER compartment. EMBO J 7: 913–918

Pellett PE, Kousoulas KG, Pereira L, Roizman B (1985) Anatomy of the herpes simplex virus 1 strain F glycoprotein B gene: primary sequence and predicted protein structure of the wild-type and of monoclonal antibody-resistant mutants. J Virol 53: 243–253

Pensiero MN, Jennings GB, Schmaljohn CS, Hay J (1988) Expression of the Hantaan virus M genome segment by using a vaccinia virus recombinant. J Virol 62: 691–702

Persson R, Wikström L, Pettersson RF (1989) Dimerization and intracellular transport of two Golgi-specific bunayavirus membrane glycoproteins. In: Compans RW, Helenius A, Oldstone MBA (eds) Cell biology of virus entry, replication, and pathogensis. Liss, New York, pp 229–241

Pesonen M, Kuismanen E, Pettersson RF (1982a) Monosaccharide sequence of protein-bound glycans of Uukuniemi virus. J Virol 41: 390–400

Pesonen M, Ronnholm R, Kuismanen E, Pettersson RF (1982b) Characterization of the oligosaccharides of Inkoo virus envelope glycoproteins. J Gen Virol 63: 425–434

Petrie BL, Graham DY, Hanssen H, Estes MK (1982) Localization of rotavirus antigens in infected cells by ultrastructural immunocytochemistry. J Gen Virol 63: 457–467

Petrie BL, Estes MK, Graham DY (1983) Effects of tunicamycin on rotavirus morphogenesis and infectivity. J Virol 46: 270–274

Petrie BL, Greenberg HB, Graham DY, Estes MK (1984) Ultrastructural localization of rotavirus antigens using colloidal gold. Virus Res 1: 133–152

Pettersson RF, von Bonsdorff C-H (1987) Bunyaviridae. In: Nermut MV, Steven Ac (eds) Animal virus structure. Elsevier, Amstredam, pp 147–157

Pettersson RF, Gahmberg N, Kuismanen E. Kääriäinen L, Rönnholm R, Saraste J (1988) Bunyavirus membrane glycoproteins as models for Golgi-specific proteins. Mod Cell Biol 6: 65–96

Pizer LI, Cohen GH, Eisenberg RJ (1980) Effect of tunicamycin on herpes simplex virus glycoproteins and infectious virus production. J Virol 34: 142–153

Poruchynsky MS, Atkinson PH (1988) Primary sequence domains required for the retention of rotavirus VP7 in the endoplasmic reticulum. J Cell Biol 107: 1697–1706

Poruchynsky MS, Tyndall C, Both GW, Sato F, Bellamy AR, Atkinson PH (1985) Deletions into an NH_2-terminal hydrophobic domain result in secretion of rotavirus VP7, a resident endoplasmic reticulum membrane glycoprotein. J Cell Biol 101: 2199–2209
Powell KFH, Gunn AR, Bellamy AR (1988) Nucleotide sequence of bovine rotavirus genomic segment 10: and RNA encoding the viral nonstructural glycoprotein. Nuclic Acids Res 16: 763
Puddington L, Machamer CE, Rose JK (1986) Cytoplasmic domains of cellular and viral integral membrane proteins substitute for the cytoplasmic domain of the VSV G protein in transport to the cell surface. J Cell Biol 102: 2147–2157
Repp R Tamura T, Boschek CB, Wege H, Schwarz RT, Niemann H (1985) The effects of processing inhibitors of N-Linked oligosaccharides on the intracellular migration of glycoprotein E2 of mouse hepatitis virus and the maturation of coronavirus particles. J Biol Chem 260: 15873–15879
Rice CM, Lenches EM, Eddy SR, Shin SJ, Sheets RL, Strauss JH (1985) Nucleotide sequence of yellow fever virus: implications for flavivirus gene expression and evolution. Science 229: 726–733
Rice CM, Strauss EG, Strauss JH (1986) Structure of the flavivirus genome. In: Schlesinger S, Schlesinger MJ (eds) The Togaviridae and Flaviviridae. Plenum, New York, pp 279–326
Richardson MA, Iwamoto A, Ikegami N, Nomoto A, Furuichi Y (1984) Nucleotide sequence of the gene encoding the serotype-specific antigen of human (Wa) rotavirus: comparison with the homologous genes from simian SA11 and UK bovine rotaviruses. J Virol 51: 860–862
Richardson WD, Roberts BL, Smith AE (1986) Nuclear location signals in polyoma virus large-T. Cell 44: 77–85
Roizman B, Batterson W (1985) Herpesviruses and their replication. In: Fields BN, Knipe DM, Chanock RM, Melnick JL, Roizman B, Shope RE (eds) Virology. Raven, New York, pp 497–526
Rönnholm R, Pettersson RF (1987) Complete nucleotide sequence of the M RNA segment of Uukuniemi virus encoding the membrane glycoproteins G1 and G2, Virology 160: 191–202
Rothman JE (1981) The Golgi apparatus: two organelles in tandem. Science 213: 1212–1219
Rottier PJ, Rose JK (1987) Coronavirus E1 glycoprotein expressed from cloned cDNA localizes in the Golgi region. J Virol 61: 2042–2045
Rottier PJM, Horzinek MC, van der Zeijst BAM (1981) Viral protein synthesis in mouse hepatitis virus strain A59-infected cells: effect of tunicamycin. J Virol 40: 350–357
Rottier P, Brandenburg D, Armstrong J, van der Zeijst B, Warren G (1984) Assembly in vitro of a spanning membrane protein of the endoplasmic reticulum: the E1 glycoprotein of coronavirus mouse hepatitis virus A59. Proc Natl Acad Sci USA 81: 1421–1425
Rottier P, Armstrong J, Meyer DI (1985) Signal recognition particle-dependent insertion of coronavirus E1, an intracellular membrane glycoprotein. J Biol Chem 260: 4648–4652
Rottier PJM, Welling GW, Welling-Webster S, Niesters HGM, Lenstra JA, van der Zeijst BAM (1986) Predicted membrane topology of the coronavirus protein E1. Biochemistry 25: 1335–1339
Sabara M, Babiuk LA, Gilchrist J, Misra V (1982) Effect of tunicamycin on rotavirus assembly and infectivity. J Virol 43: 1082–1090
Sakaguchi M, Mihara K, Sato R (1987) A short amino terminal segment of microsomal cytochrome P-450 functions both as an insertion signal and as a stop-transfer sequence. EMBO J 6: 2425–2431
Sarmiento M, Haffey M, Spear PG (1979) Membrane proteins specified by herpes simplex viruses. III. Role of glycoprotein VP7 (β_2) in virion infectivity. J Virol 29: 1149–1158
Schmaljohn CS, Patterson JL (1990) Bunyaviridae and their replication. In: Fields BN, Knipe DM, Chanock RM, Melnick JL, Roizman B, Shope RE (eds) Virology. Raven, New York, pp 1175–1194
Schmaljohn CS, Schmaljohn AL, Dalrymple JM (1987) Hantaan virus M RNA: coding strategy, nucleotide sequence, and gene order. Virology 157: 31–39
Schmidt I, Skinner M, Siddell S (1987) Nucleotide sequence of the gene encoding the surface projection glycoprotein of coronavirus MHV-JHM. J Gen Virol 68: 47–56
Shahrabadi MS, Babiuk LA, Lee PWK (1987) Further analysis of the role of calcium in rotavirus morphogenesis. Virology 158: 103–111
Shapiro D, Kos K, Brandt WE, Russell PK (1972) Membrane-bound proteins of Japanese encephalitis virus-infected chick embryo cells. Virology 48: 360–372
Shida H (1986) Nucleotide sequence of the vaccinia virus hemagglutinin gene. Virology 150: 451–462
Siddell S, Wege H, ter Meulen V (1983) The biology of coronaviruses. J Gen Virol 64: 761–776
Simons K, Fuller S (1987) The budding of enveloped viruses: a paradigm for membrane sorting? In: Burnett RM, Vogel HJ (eds) Biological organization: macromolecular interactions at high resolution. Academic, New York, pp 139–150
Simons K, Garoff H (1980) The budding mechanism of enveloped animal viruses. J Gen Virol 50: 1–21
Soler C, Musalem C, Lorono M, Espejo RT (1982) Association of virao particles and viral proteins with membranes in SA11-infected cells. J Virol 44: 983–992

Spaan W, Cavanagh D, Horzinek MC (1988) Coronaviruses: structure and genome expression. J Gen Virol 69: 2939–2952
Spear PG (1980) Herpesviruses. In: Blough HA, Tiffany JM (eds) Cell membranes and viral envelopes, vold 2. Academic, London, pp 709–750
Spear PG (1984) Clycoproteins specified by herpes simplex viruses. In: Roizman E (ed) the herpesviruses. Plenum, New York, pp 315–356
Stephens EB, Compans RW (1988) Assembly of animal viruses at cellular membranes. Annu Rev Microbiol 42: 489–516
Stern DF, Sefton BM (1982) Coronavirus proteins: structure and function of the oligosaccharides of the avian infectious bronchitis virus glycoproteins. J Virol 44: 804–812
Stirzaker SC, Both GW (1989) The signal peptide of the rotavirus glycoprotein VP7 is essential for its retention in the ER as an integral membrane protein. Cell 56: 741–747
Stirzaker SC, Whitfeld PL, Christie DL, Bellamy AB, Both GW (1987) Processing of rotavirus glycoprotein VP7: implications for the retention of the protein in the endoplasmic reticulum. J Cell Biol 105: 2897–2903
Stohlman SA, Lai MMC (1979) Phosphoproteins of murine hepatitis viruses. J Virol 32: 672–675
Stohlman SA, Wisseman CL, Eylar OR, Silverman DJ (1975) Dengue virus-induced modifications of host cell membranes. J Virol 16: 1017–1026
Strauss EG, Strauss JH (1985) Assembly of enveloped animal viruses. In: Casjens S (ed) Virus structure and assembly. Jones and Bartlett, Boston, pp 205–234
Stuve LL, Brown-Shimer S, Pachl C, Najarian R, Dina D, Bruke RL (1987) Structure and expression of the herpes simplex virus type 2 glycoprotein gB gene. J Virol 61: 326–335
Sturman LS, Holmes KV (1983) The molecular biology of coronaviruses. Adv Virus Res 28: 36–112
Sturman L, Holmes K (1985) The novel glycoproteins of coronaviruses. Trends Biochem Sci 10: 17–20
Sturman LS, Holmes KV, Behnke J (1980) Isolation of coronavirus envelope glycoproteins and interaction with the viral nucleocapsid. J Virol 33: 449–462
Sturman LS, Richard CS, Holmes KV (1985) Proteolytic cleavage of the E2 glycoprotein of murine coronavirus: activation of cell-fusing activity of virions by trypsin and separation of two different 90K cleavage fragments. J Virol 56: 904–911
Sugiyama K, Amano Y (1981) Morphological and biological properties of a new coronavirus associated with diarrhea in infant mice. Arch Virol 67: 241–251
Sullivan V, Smith GL (1987) Expression and characterization of herpes simplex virus type 1 (HSV-1) glycoprotein G (gG) by recombinant vaccinia virus: neutralization of HSV-1 infectivity with anti-gG antibody. J Gen Virol 68: 2587–25798
Sumiyoshi H, Mori C, Fuke I, Morita K, Kuhara S, Kondou J, Kikuchi Y, Nagamatu H, Igarashi A (1987) Complete nucleotide sequence of the Japanese encephalitis virus genome RNA. Virology 161: 497–510
Suzich JA, Collett MS (1988) Rift Valley fever virus M segment: cell-free transcription and translation of virus-complementary RNA. Virology 164: 478–486
Tooze J, Tooze SA (1985) Infection of AtT20 murine pituitary tumour cells by mouse hepatitis virus strain A59: virus budding is restricted to the Golgi region. Eur J Cell Biol 37: 203–212
Tooze J, Tooze SA, Warren G (1984) Replication of coronavirus MHV-A59 in sac cells: determination of the first site of budding of progeny virions. Eur J Cell Biol 33: 281–293
Tooze J, Tooze SA, Fuller SD (1987) Sorting of progeny coronavirus from condensed secretory proteins at the exit from the trans-Golgi network of AtT20 cells. J Cell Biol 105: 1215–1226
Tooze SA, Tooze J, Warren G (1988) Site of addition of N-acetyl-galactosamine to the E1 glycoprotein of mouse hepatitis virus-A59. J Cell Biol 106: 1475–1487
Torrisi MR, Cirone M, Pavan A, Zompetta C, Barile G, Frati L, Faggioni A (1989) Localization of Epstein-Barr virus envelope glycoproteins on the inner nuclear membrane of virus-producing cells. J Virol 63: 828–832
Ulmanen I, Seppälä P, Pettersson RF (1981) In vitro translation of Uukuniemi virus-specific RNAs: identification of a nonstructural protein and a precursor to the membrane glycoproteins. J Virol 37: 72–79
Vaux DJT, Helenius A, Mellman I (1988) Spike-nucleocapsid interaction in Semliki Forest virus reconstructed using network antibodies. Nature 336: 36–42
Vlasak R, Luytjes W, Leider J, Spaan W, Palese P (1988) The E3 protein of bovine coronavirus is a receptor-destroying enzyme with acetylesterase activity. J Virol 62: 4486–4690
von Bonsdorff C-H, Pettersson R (1975) Surface structure of Uukuniemi virus. J Virol 16: 1296–1307
von Bonsdorff C-H, Saikku P, Oker-Blom N (1970) Electron microscope study on the development of Uukuniemi virus. Acta Virol 14: 109–114

Vorndam AV, Trent DW (1979) Oligosaccharides of the California encephalitis viruses. Virology 95: 1–7
Ward CW, Azad AA, Dyall-Smith ML (1985) Structural homologies between RNA gene segments 10 and 11 from UK bovine, simian SA11, and human Wa rotaviruses. Virology 144: 328–336
Wasmoen TL, Kakach LT, Collett MC (1988) Rift Valley fever M segment: cellular localization of M segment-encoded proteins. Virology 166: 275–280
Webster RE, Cashman JS (1978) Morphogenesis of the filamentous single-stranded DNA phages. In: Denhardt DT, Dressler D, Ray DS (eds) The single-stranded DNA phages. Cold Spring Harbor Laboratory, Cold Spring Harbor, pp 557–569
Wengler G, Wengler G (1989) Cell-associated West Nile flavivirus is covered with E + Pre-M protein heterodimers which are destroyed and reorganized by proteolytic cleavage during virus release. J Virol 63: 2521–2526
Wenske EA, Courtney RJ (1983) Glycosylation of herpes simplex virus type I gC in the presence of tunicamycin. J Virol 46: 297–301
Westaway EG (1987) Flavivirus replication strategy. Adv Virus Res 33: 45–90
Whitfield PL, Tyndall C, Stirzaker SC, Bellamy AR, Both GW (1987) Location of sequences within rotavirus SA11 glycoprotein VP7 which direct it to the endoplasmic reticulum. Mol Cell Biol 7: 2491–2497
Wickner WT, Lodish HF (1985) Multiple mechanisms of protein insertion into and across membranes. Science 230: 400–407
Yoo D, Kang CY (1987) Nucleotide sequence of the M segment of the genomic RNA of Hantaan virus 76–118. Nuclic Acid Res 15: 6299–6300

Transport of Membrane Proteins to the Cell Surface

D. EINFELD and E. HUNTER

1 Introduction

The intracellular trafficking of molecules to discrete compartments within the cell is an absolute requirement for the normal functioning of its biosynthetic and metabolic machinery. The plasma membrane of the cell represents a unique compartment, since it is through molecules that have been specifically transported there that the cell senses its environment, transmits positive and

Department of Microbiology, University of Alabama at Birmingham, UAB Station, Birmingham, AL 35294, USA

Current Topics in Microbiology and Immunology, Vol. 170

negative stimuli, and transports in necessary precursors for growth and survival.

The general pathway for the transport of proteins to the cell surface was first described by PALADE (1975) and his colleagues and has since become the focus of intense effort in an attempt to understand the details of this vesicular transport system. The endoplasmic reticulum (ER) represents the point of entry for proteins that will traverse this complex organellar pathway. The organelles within this system include the rough and smooth ER (RER and SER), the cis-, medial-, and trans-Golgi, the trans-Golgi network (TGN), secretory vesicles and granules, and the plasma membrane (Fig. 1). Each of these organelles maintains a unique subset of resident membrane proteins that define the structure and function of that compartment. In addition, a variety of specifically targeted transitional/transport vesicles must mediate traffic between the multiple compartments of the pathway.

A molecular description of the processes which lead to the intracellular movement of a nascent polypeptide from its site of synthesis within the cytoplasm to its insertion in (or secretion from) the plasma membrane is still incomplete. Nevertheless, much progress has been made in recent years on defining the cellular machinery that is involved in this transport process and identifying the cis-acting features of proteins that modulate interactions within the pathway. This chapter deals first with the transport machinery of the cell and then with how proteins destined for transport through the system interact with it.

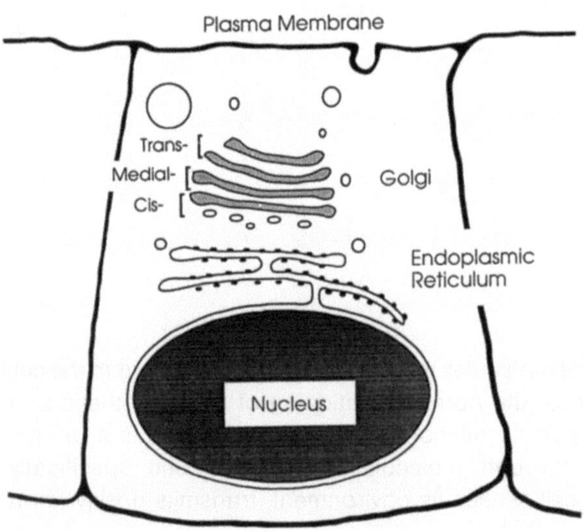

Fig. 1. Simplified representation of a eukaryotic cell showing the secretory pathway along which proteins are directed to the plasma membrane

2 The Pathway for Protein Transport to the Plasma Membrane

2.1 Translocation Across the Rough Endoplasmic Reticulum

In mammalian cells most plasma membrane proteins are cotranslationally inserted into the ER due to the presence of a signal sequence which tags the nascent polypeptide for transport through the secretory pathway of the cell (see Sect. 3.2). The cellular machinery that recognizes this sequence and directs the protein to the ER is known to involve at least two components: the signal recognition particle (SRP) and the SRP receptor (reviewed by WALTER et al. 1984). The SRP, which is a complex of six polypeptides and a 300-nucleotide RNA molecule (Fig. 2), binds to the signal sequence when the latter is exposed on the ribosome during translation and temporarily arrests elongation of the protein. This interaction apparently involves the 54-kDa polypeptide, since it can be cross-linked to the signal sequence (KRIEG et al. 1986; KURZCHALIA et al. 1986) and SRP reconstituted with modified 54-kDa polypeptides have reduced binding affinity for the signal peptide (SIEGEL and WALTER 1988b). Elongation arrest appears to be mediated by two associated polypeptides of 9 and 14-kDa, since reconstituted SRPs lacking these proteins are defective in this property (SIEGEL and WALTER 1988b).

Targeting of the nascent polypeptide to the ER requires an interaction between the SRP 68/72kD protein (SIEGEL and WALTER 1988b) and its ER-associated receptor, a heterodimeric molecule also called "docking protein". Binding of SRP by its receptor results in dissociation of SRP from the nascent polypeptide, continuation of translation, and translocation of the polypeptide across the ER membrane (GILMORE and BLOBEL 1983; GILMORE et al. 1982; MEYER

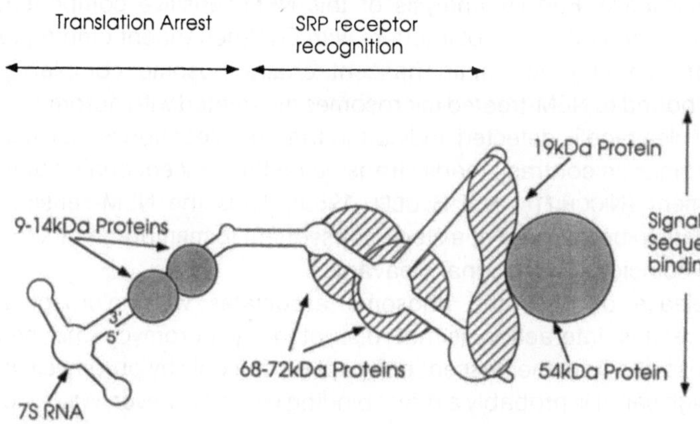

Fig. 2. Model for mammalian signal recognition particle (SRP), consisting of a 300-nucleotide 7S RNA and is proteins: P9, P14, P19, P54, P68, and P72. All proteins except P54 bind to the RNA; P54 binds to P19

et al. 1982). Following interaction with SRP and the SRP receptor, the nascent polypeptide is engaged in a guanosine triphosphate (GTP)-dependent association with microsomes that is resistant to disruption by ethylenediaminetetraacetic acid or high salt (CONNOLLY and GILMORE 1986). Nonhydrolyzable analogs of GTP can substitute for the latter in formation of this complex, while guanosine diphosphate is a potent inhibitor. Subsequent experiments indicate that in the presence of SRP receptor GTP allows dissociation of SRP from nascent chain-ribosome complexes (CONNOLLY and GILMORE 1989). The α chain of the SRP receptor is capable of binding GTP and possesses some of the consensus sequences of GTP-binding proteins (CONNOLLY and GILMORE 1989).

Additional cellular factors have been implicated in targeting membrane proteins to the ER and in translocation across this membrane, but the precise roles of these factors have not been defined. Signal sequences have been cross-linked to integral ER membrane proteins of 35 kDa and 42 kDa (ROBINSON et al. 1987; WIEDMANN et al. 1987), and interaction of SRP-arrested peptides with this 35-kDa signal sequence receptor was dependent on SRP receptor, presumably for dissociation of SRP from the signal sequence. Thus it is likely that this 35-kDa molecule is involved at a step later than SRP recognition.

Alkylation of an ER component, distinct from the docking protein, with *N*-ethylmaleimide (NEM) can render microsomes inactive in translocation (HORTSCH et al. 1986). A cytoplasmic fragment composed of the C-terminal three quarters of SRP receptor α chain is released by protease treatment of microsomes but is able to reassociate with the microsomes and restore their SRP receptor activity (GILMORE et al. 1982; HORTSCH et al. 1985). When microsomes stripped of this component of SRP receptor are treated with NEM they retain the ability to reassociate with the cytoplasmic α fragment but translocation activity is not restored. Under these conditions the SRP receptor appears functional in that the reconstituted microsomes can relieve SRP-mediated translocation arrest. The alkylated microsomes retain the ability to bind ribosomes and signal peptidase remains active. Further analysis of this NEM-sensitive component indicates that its inactivation does not interfere with GTP-dependent binding to microsomes of truncated preprolactin nascent chain-ribosome complexes. When complexes bound to NEM-treated microsomes are treated with puromycin, no signal peptide cleavage is detected, indicating that translocation is blocked. Untreated microsomes, in contrast, readily translocate the nascent chains after puromycin treatment (NICCHITTA and BLOBEL 1989). Thus the NEM-sensitive component appears to be involved at a stage subsequent to membrane binding but preceding or coincident with signal cleavage.

Following release of SRP, the ribosome associates with microsomal membranes. Since this interaction is not disrupted by puromycin-induced release of the ribosomes from the nascent polypeptide, but only by puromycin in the presence of high salt, it is probably a direct binding event. However, WILSON et al. (1988) have reported that despite puromycin treatment, nascent polypeptides can remain largely associated with the ribosome. The molecular basis of the interaction remains unresolved but appears to involve an ER component that is protease sensitive and distinct from the ribophorins (HORTSCH et al. 1986). The posttranslational translocation of polypeptides or N-terminal fragments into mammalian microsomes in vitro exhibits a dependence on the association of the peptide with ribosomes (CHAO et al. 1987; MUECKLER and LODISH 1986; PERARA et al. 1986). The ribosome may thus maintain the nascent polypeptide in an unfolded state compatible with translocation (PERARA et al. 1986). However, there are size limits beyond which postranslational translocation does not occur (AINGER and MAYER 1986; CHAO et al. 1987; PERARA et al. 1986; SIEGEL and WALTER 1985). This size limitation is also a factor in the association of the nascent chains with microsomes (SIEGEL and WALTER 1988a), and thus can reflect the need for an accessible signal sequence as well as translocation competence. Since translocation generally appears to be cotranslational in mammalian cells the in vivo significance of this function of the ribosome remains to be determined.

A variety of interesting mutants defective in protein translocation into the ER have been identified in yeast. Although the secretory pathways of yeast and mammalian cells exhibit differences both in the architecture of the transport system and in the diversity of mechanisms of translocation, some functions are clearly shared between the two systems. We will therefore briefly discuss the phenotypes of some of the mutants here. Unlike in vitro mammalian systems, yeast systems can carry out posttranslational translocation of fully elongated proteins released from ribosomes. This translocation is dependent on the presence of a signal sequence and the availability of adenosine triphosphate (ATP) as an energy source, but is limited to a subset of secretory proteins (ROTHBLATT et al. 1987). Yeast mutants deficient in the expression of hsp70 heat shock proteins accumulate a precursor form of prepro-α factor, a protein that is translocated posttranslationally in vitro, and to a lesser extent the precursor of invertase, a protein that is translocated cotranslationally in vitro (Deshaies et al. 1988). Identification of an hsp70 protein as a factor in yeast postribosomal supernatant that promotes the in vitro translocation of prepro-α factor points to a chaperone-like role for this heat shock protein in maintaining translocation competence of the nascent polypeptide (possibly by preventing folding of the protein to its mature conformation or by directing unfolding of a prematurely folded polypeptide) (CHIRICO et al. 1988). As noted above, the maintenance of translocation competence may be part of the function of the ribosome in mammalian cells. A translocation-promoting activity in yeast cytosol that differs from that of hsp70 in being sensitive to NEM inactivation has also been reported (WATERS et al. 1986), but the molecule or molecules involved have not been identified.

Yeast *sec61* mutants are temperature sensitive for growth and accumulate unmodified, and thus untranslocated, precursors of secreted proteins at both nonpermissive and permissive temperatures, although full expression of the translocation defect is only observed at 37 °C (DESHAIES and SCHEKMAN 1987). The accumulated precursors remain exposed to the cytoplasm: they are accessible to protease in spheroplast lysates, are not glycosylated, and apparently retain their signal sequences. The block in translocation, however, occurs after membrane binding, since these molecules copurify with the membranes in fractionation experiments. Employing a genetic selection scheme similar to that which yielded *sec61*, a distinct translocation mutation, *ptll*, has been identified (TOYN et al. 1988). Like *sec61*, this mutation causes accumulation of secretory protein precursors in vivo that are susceptible to exogenous protease digestion. While it is not clear whether these precursors are membrane bound, microsomal membranes have been implicated as the site where the *ptll* lesion is expressed. The accumulated precursors can be translocated when the *ptll* mutants are returned to permissive temperature, indicating that posttranslational translocation can occur in vivo. Thus it seems likely that in yeast as well as in mammalian cells, a series of events following association of the nascent polypeptide to the ER is required for successful translocation across the membrane itself. It remains to be seen whether the yeast factors discussed here will lead to the identification of equivalent mammalian components or whether distinct processes have evolved in the two systems. An understanding of the actual mechanisms involved in translocating the protein through the membrane remains elusive, as does the nature of the environment (protein channel vs modified lipid bilayer) through which the protein is moved.

Widely diverse organisms employ precursor proteins lacking the folded conformation of the mature protein as a substrate for translocation of these proteins across membranes. In this aspect translocation of proteins across the ER resembles protein import into mitochondria and export of proteins across the cytoplasmic membrane of bacteria (*Escherichia coli*). The conformation of these precursors is generally described as "unfolded" due to lack of structural features of the mature protein, increased sensitivity to proteases, or stimulatory effects of denaturants such as urea on translocation of mature proteins. Several cytoplasmic proteins have been identified in *E. coli* that promote protein export. These factors, including SecB, trigger factor, GroEL, and GroES, appear to maintain precursor proteins in a translocation-competent conformation (LECKER et al. 1989; TURN et al. 1988; WATANABE and BLOBEL 1989; WEISS et al. 1988). Some differences among these factors in specificity for precursor proteins are apparent. An analogous function is performed by hsp70 proteins, specifically the products of SSAI and SSA2, in yeast. While GroEL and GroES are heat shock proteins of *E. coli*, the yeast proteins appear to belong to a different structural class. The yeast homolog of GroEL appears to be a mitochondrial hsp60 involved in folding and assembly of proteins within this organelle (CHENG et al. 1989; READING et al. 1989). Hsp70 proteins have not been shown to be required for protein translocation in mammalian cells, but the cotranslational nature of translocation in these cells may preempt a requirement for this function. While analysis of a posttranslational system might be used to examine the role of mammalian hsp70 proteins, as suggested by DESHAIES et al. (1988), the fact that substrates for these systems cannot exceed a certain length supports the view that hsp70 proteins might at most play a limited role. Thus there appears to be a diversity of factors employed by various organisms in conforming protein precursors to the constraints of translocation. It remains to be seen to what extent this picture reflects an incomplete characterization of these factors or indicates true diversity.

2.2 Protein Modifications in the Endoplasmic Reticulum and the Development of Transport Competence

During and after the process of translocation itself, components of the ER play a variety of roles in catalyzing posttranslational modifications such as proteolysis and glycosylation of the extruded protein. In addition it has become increasingly clear that molecules resident within the lumen of the ER assist in the correct folding of translocated (presumably transiently denatured) polypeptides and their assembly into oligomeric complexes. The attainment of a correct structure appears to be critical for transport out of the ER to the Golgi, and may in fact be the rate-limiting step in this process.

2.2.1 Proteolysis/Signal Peptidase

After translocation of a nascent chain across the ER has been initiated, the signal peptide, which marked the protein for export, is removed from most proteins by proteolytic cleavage. Signal peptidase is an integral membrane protein and has been studied using detergent-solubilized enzyme from dog pancreas (JACKSON and WHITE 1981) and hen oviduct (LIVELY and WALSH 1983). The canine signal peptidase has been purified as a complex yielding six polypeptide bands in sodium dodecyl sulfate–polyacrylamide gel electrophoresis (EVANS et al. 1986). Two of these proteins appear to be differently glycosylated forms of a single polypeptide (SHELNESS et al. 1988). Hen oviduct signal peptidase is composed of two protein subunits, one of which is glycosylated (BAKER and LIVELY 1987). The glycoprotein components of these complexes are similar in size, and the canine cDNA clone indicates amino acid sequence similarity between the encoded glycoprotein and tryptic fragments from the avian glycoprotein (SHELNESS et al. 1988). Signal peptidase activity in bacteria is catalyzed by single polypeptide enzymes. Two classes of signal peptidases have been described for *E. coli*, signal peptidase I (SPase I) and signal peptidase II (SPase II). The former has been cloned into pBR322 (DATE and WICKNER 1981) and has been shown to accurately cleave eukaryotic precursor proteins as well as bacterial protein precursors (TALMADGE et al. 1980). Conversely, the eukaryotic signal peptidase will accurately cleave prokaryotic proteins (WATTS et al. 1983). The functions of the additional polypeptides in the mammalian enzyme complex have not been determined. With apparent molecular weights of 25 kDa or less, all of the polypeptides are smaller than the translocation factors discussed above. Yeast *sec11* mutants appear to be specifically defective in signal peptide cleavage, but the *SEC11* gene shows no homology to bacterial leader peptidase (BOEHNI et al. 1988). Whether the predicted 19-kDa yeast gene product is the functional homolog of a component of the complex seen in higher eukaryotes remains to be seen. As yet the yeast signal peptidase complex has not been characterized.

Cleavage of the signal peptide is not essential for translocation to be completed, since mutations within the signal peptidase recognition site can block the former without affecting the latter (NOTHWEHR et al. 1989). However, proteins in which a normally cleaved signal peptide is not removed may be unable to be transported out of the ER.

2.2.2 Addition of Heterologous Components to the Polypeptide

The ER contains the enzymatic machinery for transferring the core oligosaccharide chains acquired by *N*-glycosylated proteins. The $Glc_3Man_9GlcNAc_2$ unit is transferred in a single step from a membrane lipid anchor to luminal asparagine residues of the nascent polypeptide that occur in the sequence motif Asn-X-Ser or Asn-X-Thr. The oligosaccharyl transferase consists of an integral membrane component which functions together with a luminal protein with binding specificity for the glycosylation sequence (GEETHA-HABIB et al. 1988). This luminal protein apparently has multiple functions (see below). Also present in the ER are a pair of glucosidases and an α-mannosidase that remove the three glucose residues and zero, one, or three mannose residues (for review see KORNFELD and KORNFELD 1985).

Additional modifications occur within the lumen of the ER. For example, addition of the C-terminally attached glycosyl-phosphatidylinositol moiety by which some plasma membrane proteins are anchored in the membrane occurs within the ER (FERGUSON and WILLIAMS 1988; LOW and SALTIEL 1988). Attachment of the glycolipid is preceded by or occurs in concert with removal of a C-terminal segment that includes 10–20 hydrophobic residues. The signal for this modification may involve sequences N-terminal to the cleavage site in addition to the hydrophobic sequence itself (CARAS and WEDDELL 1989; CARAS et al. 1989).

Plasma membrane proteins can also be modified by attachment of the fatty acid, palmitic acid, typically via a thioester linkage to cysteine residues. This phenomenon has been the subject of recent reviews (SCHULTZ et al. 1988; TOWLER et al. 1988). Palmitate may be added to different domains of an integral membrane protein. In the vesicular stomatitis virus (VSV) G protein (ROSE et al. 1984) and the transferrin receptor (JING and TROWBRIDGE 1987) modification of the cytoplasmic domain is observed, whereas in other proteins such as the MHC antigens (KAUFMAN et al. 1984) it can occur at a site within the membrane-spanning anchor domain. Palmitylation is not unique to integral membrane proteins, nor does there appear to be a unique cellular site where this modification occurs. For example, transferrin receptor, myelin lipophilin, and bovine rhodopsin are apparently palmitylated in the late Golgi or at the plasma membrane, and proteins synthesized in the cytosol, such as $p21^{ras}$, can also be palmitylated. Other proteins such as VSV G protein appear to be acylated while still at a late ER or pre-Golgi site, and in an endoglycosidase H (endo H)-sensitive form (SCHMIDT and SCHLESSINGER 1980). But palmitylation of the ts045 mutant of

VSV G protein is blocked at the non-permissive temperature at which the mutant protein is unable to exit the ER (SCHMIDT and SCHLESSINGER 1979). These observations suggest that in at least some cases palmitylation may be an early modification and occur in the ER, but that the nascent protein must progress beyond the translocation site. The acyltransferase has been only partially purified, but an in vitro activity assay indicates a high level of activity associated with the RER (BERGER and SCHMIDT 1985; MACK et al. 1987).

2.2.3 Association with Protein Chaperones/Protein Folding

There is growing evidence that a group of polypeptide chain binding (PCB) proteins, or molecular chaperones (ELLIS 1987; ELLIS et al. 1989), play crucial roles in the processes of protein folding in both prokaryotic and eukaryotic cells (ROTHMAN 1989). A resident ER protein called BiP is identical to the glucose-regulated protein GRP78. It can be coimmunoprecipitated with newly synthesized immunoglobulin heavy chains and influenza hemagglutinin (HA) polypeptides, molecules that form part of multimeric complexes (BOLE et al. 1986; COPELAND et al. 1987; GETHING et al. 1986). This association appears to be transient for transported wild-type proteins but persists with mutant proteins that are retained in the ER. Binding of BiP to subunits of oligomeric proteins can be dissociated by an ATP-dependent mechanism (MUNRO and PELHAM 1986). While a variety of functions have been proposed for BiP, it seems likely that it plays a role in regulating the correct folding of newly translocated molecules within the lumen of the ER (ROTHMAN 1989). Expression of BiP in mammalian cells is elevated by accumulation of glycoprotein precursors in the ER (KOZUTSUMI et al. 1988). Accumulation of the precursors in the ER was achieved using mutant proteins whose transport was blocked in the ER, as well as by glycosylation inhibitors that affected transport of wild-type proteins. These precursors, then, would appear to be targets for interaction with BiP. The yeast homolog of BiP, however, which is encoded by a single gene (*KAR2*) and is required for viability, is induced not only by abnormal precursors within the ER but also by accumulation within the ER of normally folded core glycosylated proteins in mutant cells defective for transport out of the ER (NORMINGTON et al. 1989; ROSE et al. 1989). Understanding the regulation of BiP should provide further insights into the processing of proteins within the ER. It remains to be determined whether additional PCB proteins operate within the ER and whether proteins will differ in their requirements for such catalysts.

Protein disulfide isomerase (PDI) is also a resident of the ER lumen and capable of catalyzing disulfide bond formation and reduction. Formation of correct disulfide bonds is essential to the acquisition of native conformation by cysteine-containing proteins. Microsomes depleted of PDI by exposure to pH9 exhibit a defect in formation of intrachain disulfide bonds in newly translocated proteins (BULLEID and FREEDMAN 1988). While the treated microsomes are most likely depleted of many luminal ER proteins, microsomes reconstituted with purified PDI regain ability to catalyze intrachain disulfide bond formation.

Interestingly, PDI appears to be a multifunctional protein (FREEDMAN 1989) which is most likely identical to the glycosylation site binding protein involved in *N*-linked glycosylation (GEETHA-HABIB et al. 1988).

2.3 Transport from Rough Endoplasmic Reticulum to Cis-Golgi

Movement of membrane proteins from the ER to the plasma membrane can be described by a model of directed transit through a series of membranous structures that includes the compartments of the Golgi complex and culminates in fusion of secretory/transport vesicles with the plasma membrane. The morphologically and biochemically distinct structures such as the ER and the cisternae of the Golgi are linked by transport vesicles that permit exocytic protein movement via membrane fusion. Such a system, however, requires a mechanism for maintaining the integrity and distinctness of the linked organelles. The distinct character of the intermediate organelles makes possible directed movement of proteins to the plasma membrane, provided that at each step transport vesicles have a way to specifically recognize the appropriate target membrane. Understanding how such a system distinguishes transient proteins from the resident proteins at each intermediate stage, as well as the biochemical mechanisms of vesicle formation, targeting, and fusion, remains a goal of ongoing research.

Proteins that transit the ER must move from the translocation site to regions of the ER that give rise to transport vesicles. Many of the factors directing protein translocation are found to be selectively restricted to the RER, but other proteins are distributed throughout the rough and smooth domains of the ER (for review see ROSE and DOMS 1988). Itinerant membrane proteins ultimately appear in smooth membrane structures that bud from the ER and have been identified as an intermediate stage in transport from the ER to the Golgi.

Membrane glycoproteins normally targeted to the plasma membrane accumulate in a post-ER, pre-Golgi structure in cells incubated at 15 °C (BALCH et al. 1986; COPELAND et al. 1988; SARASTE and KUIMANEN 1984; SARASTE et al. 1986). The glycoproteins blocked at this stage of transport remain endo H sensitive, and thus have not been exposed to the oligosaccharide trimming enzymes found in the cis-Golgi (COPELAND et al. 1988; SARASTE and KUIMANEN 1984). Under these conditions Semliki Forest virus glycoproteins are concentrated in vesicles that appear to bud from the ER and are also detected in large vacuoles in the vicinity of the Golgi or in tubular extensions of these vacuoles (SARASTE and KUIMANEN 1984). The reduced temperature apparently disrupts transport by blocking fusion of the transport intermediates with the Golgi or interfering with an earlier step that is a prerequisite for fusion. At 15 °C the Golgi also exhibits a reduced number of cisternae and a more vesicular composition (SARASTE and KUIMANEN 1984), probably as a consequence of the transport block. The intermediate observed at 15 °C is most likely an expansion of the transitional element observed by PALADE (1975) to be a morphologically distinct extension of the ER proximal to the cis-Golgi. This intermediate structure, located in the region of transitional elements on the cis side of the Golgi, has been further identified as the site of budding of the coronavirus MHV-A59 (MAYER et al. 1988).

Vesicles mediating ER to Golgi transport have been reported from the hepatoma cell line HepG2 based on density gradient analysis of pulse-labeled cells (LODISH et al. 1987). These vesicles have a lighter density than ER or Golgi membranes and carry endo H-sensitive glycoproteins that retain oligosaccharide side chains characteristic of the ER (LODISH et al. 1987). The entry of secretory proteins into these vesicles is consistent with observations on transport of proteins from the ER. The rates at which various proteins appear in the vesicles reflect the relative rates at which they acquire endo H resistance. Moreover, proteins that are inhibited in export from the ER by 1-deoxynojirimycin are also specifically blocked, in the presence of the inhibitor, from entering these vesicles.

The exit of proteins from the ER is prevented by inhibitors and uncouplers of oxidative phosphorylation. ATP is required for proteins to move from the ER to the pre-Golgi vesicular intermediate observed at 15 °C (BALCH et al. 1986). An additional requirement for progression to this transitional body is revealed by viral glycoproteins that are temperature sensitive for exit from the ER. The ts-1 mutant of SFV glycoprotein and the ts045 mutant of VSV G protein accumulate in the pre-Golgi compartment when cells are incubated at 15 °C, and are fully transport competent when the cells are subsequently shifted to the restrictive temperature (BALCH et al. 1986; SARASTE and KUIMANEN 1984). In contrast, at the restrictive temperature these mutant proteins do not leave the ER. Thus the temperature-sensitive lesion, which is known to affect oligomerization, renders the proteins incompatible for transport to this intermediate compartment but does not interfere with transport beyond it. The connection between the failure to oligomerize and the block in transport is not clear. Movement of proteins from the ER to this intermediate compartment, then, not only requires ATP for vesicle formation but also exhibits selectivity for what proteins are included in the vesicles. The selectivity presumably accounts for the variable rates at which proteins exit the ER, a step that is rate-limiting for the transport of most proteins. It is still unknown whether this selectivity operates directly during vesicle formation or whether budding occurs at specific sites within the ER that exhibit restricted access.

Transport between the ER and the Golgi complex has also been investigated using the fungal product brefeldin A (BFA). At lower concentrations (1 μg/ml) this lipophilic molecule retards secretion, apparently acting specifically on transport from the ER to the Golgi (MISUMI et al. 1986). At an intermediate concentration (2.5 μg/ml) the Golgi is disassembled, while at 10 μg/ml transport is completely blocked and morphological changes include dilation of the ER as well as loss of the Golgi complex (FUJIWARA et al. 1988; MISUMI et al. 1986). The effects of different concentrations appear to be due to degradation of the drug during incubation with the cells (FUJIWARA et al. 1988).

Analyses of two different systems at 10 μg/ml BFA has led to the conclusion that this reagent not only blocks transport from the ER but also leads to the appearance of cis and medial-Golgi components within the ER, so that endo H-resistant oligosaccharides are found associated with proteins in the ER (DOMS et al. 1989; LIPPINCOTT-SCHWARTZ et al. 1989). Results from additional experiments suggest that BFA can also cause movement back to the ER of proteins that are not permanent residents of the cis/medial-Golgi (DOMS et al. 1989). Reverse transport of vesicles from the Golgi to the ER is probably necessary for maintenance of the two organelles in the context of exocytic transport, but under normal conditions Golgi-specific components would be excluded from the recycling vesicles. Since the effects of BFA are reversible, it would be interesting to determine the fate of putative transport-regulating molecules during distortion and self-correction of the transport system.

A number of factors involved in ER to cis-Golgi transport have been identified in an in vitro system that employs semiintact cells (BECKERS et al. 1987). Transport is detected by the acquisition of the endoglycosidase D-sensitive $Man_5GlcNAc_2$ species of oligosaccharide that is formed upon arrival in the cis-Golgi. Detection is enhanced using a mutant of VSV G protein whose transport out of the ER is temperature sensitive and, as the host cell, the 15B CHO mutant which lacks an enzyme required for processing the oligosaccharide beyond this stage. Transport requires ATP and cytosolic factors. A membrane-bound component has been implicated as the target of irreversible inhibition of transport by the nucleotide analog GTPγS (BECKERS and BALCH 1989). This factor is thought to be a G protein based on its sensitivity GTPγS and AlF_4^-. Transport in this system is also sensitive to the Ca^{2+} chelator ethyleneglycoltetraacetic acid (EGTA), which inhibits reversibly and at a step subsequent to GTPγS inhibition (BECKERS and BALCH 1989). A monoclonal antibody to the NEM-sensitive factor (NSF) originally identified as a required component for intercompartmental transport within the Golgi complex (BLOCK et al. 1988) inhibits transport from the ER to the Golgi (BECKERS et al. 1989). NSF is required at a step concurrent with or subsequent to the Ca^{2+}-requiring step. The yeast homolog of NSF is encoded by *SEC18* (WILSON et al. 1989), and mutants at this locus are also defective in transport from the ER. An additional factor required for transport in yeast is the *SEC23* gene product, which must be present on the acceptor Golgi membrane (RUOHOLA et al. 1988).

2.4 Intra-Golgi Transport

2.4.1 Structure of the Golgi Complex

The Golgi complex is a set of subcompartments that comprises at least three biochemically distinct units (reviewed by DUNPHY and ROTHMAN 1985). A prominent feature of the Golgi is the presence of enzymes that transform oligosaccharide chains into their mature form. The sequential action of these enzymes and the correlation of this with the organization of the Golgi has been reviewed (KORNFELD and KORNFELD 1985). The cis-Golgi contains a mannosidase that trims *N*-linked oligosaccharides, and enzymes that attach targeting mannose-6-phosphate residues to lysosomal enzymes. In the medial-Golgi the mannose core is trimmed further and *N*-acetylglucosamine and fucose residues are added. Galactose and sialic acid residues are acquired in the trans-Golgi. The subdivision of the Golgi may contribute to the processing of oligosaccharides by compensating for lack of strick substrate specificity among the various glycosyltransferases and glycosidases. Compartmentalization may be important for other processes such as proteolytic cleavage/activation of proteins, allowing cleavage to occur only at a specific stage of maturation within the Golgi. Lysosomal enzymes are marked for sorting by phosphorylation within the cis-Golgi; however, they are apparently not diverted from the exocytic pathway before the trans-Golgi. Sorting of proteins into secretory granules and secretory vesicles appears to take place in the TGN.

2.4.2 Vesicular Transport Within the Golgi Complex

Transport between Golgi compartments is a vesicular process that can be reconstituted in vitro when Golgi membranes are incubated in the presence of ATP and cytosol. Like ER to Golgi transport, this process can be inhibited by GTPγS and requires NSF. NSF associates with Golgi membranes but is also free in the cytosol. During the process of budding, vesicles acquire a coat distinct from clathrin. Removal of this coat appears to be an early step following interaction of the vesicle with the target membrane. In the presence of GTPγS, coated vesicles accumulate in the in vitro system (MELANCON et al. 1987). GTPγS exerts its inhibitory effect specifically on acceptor Golgi membranes and not on the donor membranes, implying that the target of this nucleotide analog, presumably a GTP-binding protein, plays a role in the interaction between the target membrane and vesicles rather than in vesicle formation (MELANCON et al. 1987). When NSF is inactivated by NEM, uncoated vesicles accumulate (MALHOTRA et al. 1988). That these uncoated vesicles have arisen from coated vesicles is indicated by order of addition experiments which reveal that NEM inhibits transport at a step later than does GTPγS (ORCI et al. 1989). The NSF-dependent step occurs after interaction between vesicles and target membranes to form a prefusion complex, a step distinct from the actual fusion (MALHOTRA et al. 1988). The accumulation of uncoated vesicles seen in the absence of functional NSF may be due to failure to form this complex. While much of this information was derived from a system detecting transport from the cis-Golgi to the medial-Golgi, the same features appear to be true of transport to the trans-Golgi (ROTHMAN 1987b).

Additional evidence for the involvement of GTP-binding proteins in the exocytic pathway comes from yeast secretion mutants. Mutation at the *SEC4* gene causes accumulation of post-Golgi vesicles. This gene encodes a protein having sequence homology to ras proteins and the yeast GTP-binding protein YPT1 (SALMINEN and NOVICK 1987). SEC4 and YPT1 are most similar at those sequences implicated in binding and hydrolysis of GTP. It is not clear whether the *SEC4* gene product is a component of the transport vesicles, the plasma membrane, or an accessory molecule regulating interaction between these two. A temperature-sensitive mutant at the *YPT1* locus (*ypt1-1*) exhibits abnormal cytoplasmic membrane structures and a defect in invertase secretion at the nonpermissive temperature (SEGEV et al. 1988). In addition intracellular accumulation of incompletely glycosylated molecules occurs in this mutant regardless of temperature. A similar secretion defect and cytoplasmic membrane accumulation have been reported under restrictive conditions for a temperature-sensitive revertant (*ypt1*^{ts}) of a lethal *YPT1*mutant (SCHMITT et al. 1988). The immunofluorescent staining pattern of the YPT1 protein correlates with Golgi membranes and the antibodies appear specific for Golgi in mouse cells (SEGEV et al. 1988). In yeast the YPTI protein is also localized to small buds, the site of protein secretion in yeast. Thus studies with yeast support the view that GTP-binding proteins are involved in the exocytic transport system; however, since additional effects are seen on cell growth, sporulation, and calcium uptake in the *YPT1* mutants (SCHMITT et al. 1988; SEGEV and BOTSTEIN 1987), the function of these proteins is not clear.

2.5 Transport from the Trans-Golgi Network to the Plasma Membrane

While a majority of membrane-spanning proteins are transported from the Golgi to the plasma membrane in a constitutive, unregulated fashion, secretion of soluble proteins involves at least two pathways between the Golgi and the plasma membrane. The proteins entering the regulated pathway may be sequestered for long periods in secretory granules that are transported to and fuse with the plasma membrane in response to specific stimulatory signals. Immunocytochemical analyses indicate that sorting into the constitutive and regulated pathways occurs in the most distal region of the Golgi (ORCI et al. 1987). The cytochemical characteristics of this compartment, acid phosphatase positive and thiamine pyrophosphatase negative, match those of the TGN defined by GRIFFITHS and SIMON (1986). The TGN is distinguished from the trans-Golgi cisterna by virtue of its more reticular structure and the absence of thiamine pyrophosphatase (GRIFFITHS and SIMONS 1986). Presumably the TGN is the structure that is elaborated when exocytic transport is blocked by incubation of cells at 20 °C. Under these conditions viral glycoproteins normally transported to the plasma membrane are observed in vacuolar structures that are associated with morphologically normal Golgi stacks (FULLER et al. 1985; GRIFFITHS et al. 1985; MATLIN and SIMONS 1983). The oligosaccharides of the accumulated proteins indicate exposure to the glycosylation enzymes of the Golgi, including the trans cisternae. When the temperature is raised these proteins move to the cell surface without detectable movement into the Golgi (SARASTE and KUIMANEN 1984).

Membrane proteins transported to the plasma membrane in a constitutive fashion are detected in noncoated vesicles morphologically distinct from the granules carrying regulated secretory molecules. Formation of secretory granules involves condensation of their contents followed by budding of membrane-bound structures coated with clathrin (ORCI et al. 1985). Maturation of the secretory granules is marked by condensation of a dense core and loss of the coat. Discrimination between these classes of proteins may affect their access to subdomains of the TGN, since the constitutively secreted HA of influenza virus is not detected in the membrane at the site of budding granules (ORCI et al. 1987).

Analyses of fusion proteins have given results consistent with a model whereby proteins enter the constitutive pathway by default but require a positive signal to be directed to the regulated pathway (MOORE and KELLY 1986). A set of 25-kDa proteins detected in crude Golgi membrane preparations from canine pancreas has been identified by their ability to bind specifically to the regulated secretory proteins prolactin and insulin, but not to proteins such as immunoglobulin or albumin which are secreted constitutively (CHUNG et al. 1989). Functionally and immunologically related proteins of similar size are also present in mouse pituitary cells (line AtT-20 and, following hormone induction of storage granule formation, pituitary GH3 cells). Binding of secreted proteins to these molecules occurs at neutral pH but is abolished at pH5. These proteins are attractive candidates for components for a sorting system, but their membrane association and topology have not been defined and their possible function in such a system may depend on additional factors. These proteins appear capable of multivalent binding of a variety of regulated secretory proteins, and its has been suggested they might play a role in granule formation by aggregating regulated secretory molecules (CHUNG et al. 1989).

Perforated MDCK cells release viral glycoproteins in a form expected for transport vesicles released from the Golgi (BENNETT et al. 1988). These proteins appear to be present in vesicles based on sedimentation analyses and exhibit the appropriate membrane topology. The vesicles show specificity for exocytic proteins; Golgi markers are released at lower levels, and plasma membrane proteins exhibit a different sedimentation behavior. Release of the Golgi vesicles is stimulated by ATP and the Ca^{2+} chelator EGTA. The differential effect of EGTA here is striking in comparison with its inhibitory effect on ER to Golgi transport.

Transport vesicles function at multiple stages of the exocytic pathway as links between organellar structures. A common feature of the interaction of these vesicles with their target membranes is the use of GTP-binding proteins and NSF. While this conservation is not surprising in light of the apparent complexity of this process, it leaves unanswered the question of what determines the specificity and directionality of transport. NSF may play a general role in establishing a fusogenic complex, but its precise role and the identity of ligands or receptors are as yet not clear.

The process by which these transport vesicles arise is not known. ATP and cytosolic components are required for budding of vesicles, but the initiating factors and the process of vesicle formation remain poorly understood. In the case of transport vesicles originating within the Golgi, the nonclathrin coat is found specifically associated with the bud as it is formed and is absent from the adjoining membrane (MALHOTRA et al. 1988; ORCI et al. 1989), suggesting that coat assembly and bud formation might be directly connected. It seems likely that the vesicles arise from specific domains of the membranes of the ER and Golgi cisternae and that access to these domains determines the content of the budded vesicles. However, the factors that have been identified to date provide only an incomplete picture of what is involved in this process.

3 Cis-Acting Features of Proteins Involved in the Transport Process

3.1 Bulk Flow Vs Directed Transport of Proteins in the Transport Pathway

An initial view of the transport process postulated the involvement of "sorting signals", located within the structure of the newly synthesized proteins, that would interact with membrane-bound receptors in the RER and Golgi apparatus of the cell and actively direct the molecule through the secretory pathway (for review see SABATINI et al. 1982; SILHAVY et al. 1983). This concept was supported by the fact that cell, as well as viral, glycoproteins could be targeted to specific points in the secretory pathway, and that cells could transport and secrete a variety of glycosylated and unglycosylated proteins at distinctly different rates (FITTING and KABAT 1982; GUMBINER and KELLY 1982; KELLY 1985; LEDFORD and DAVIS 1983; LODISH et al. 1983; STROUS and LODISH 1980).

An alternative concept of protein movement through the secretory pathway has been expounded by ROTHMAN (1987a) who proposes that secretory and membrane proteins could be carried from the ER and Golgi in the lumen and walls of unselective transport vesicles in a rapid process of "bulk" flow. The distribution and targeting of proteins through the secretory pathway would thus be specified by the presence of "retention signals" which would retain these molecules in discrete compartments. Three lines of evidence have been used to support this bulk flow/retention concept: (1) the identification of signals that specify retention, (2) experiments suggesting that the rate of bulk flow may be as fast as the fastest rate of secretion, and (3) the identification of populations of transport vesicles formed at the ER and Golgi that could serve as unselective bulk carriers.

The first evidence indicating that signals might specify retention within the ER came from a mutational analysis of the rotavirus VP7 glycoprotein (PORUCHYNSKY et al. 1985). Deletions within one of the hydrophobic domains of this 38-kDa protein resulted in its secretion into the culture medium, which was interpreted to mean that a retention signal normally prevented its export from the ER. More recently, PAABO et al. (1987) showed that truncation of the cytoplasmic domain of the adenovirus E19 glycoprotein, normally retained as an integral membrane protein in the ER, resulted in its transport to and insertion in the plasma membrane. And MACHAMER and ROSE (1987) showed that certain deletions within the coronavirus E1 glycoprotein, normally a resident of the Golgi, resulted in its transport out of that organelle to the plasma membrane. Thus, in each of these examples removal of peptide sequences resulted in a loss of retention within the secretory pathway and redirection to the plasma membrane.

MUNRO and PELHAM (1987) first described a discrete retention signal that was responsible for keeping soluble proteins within the ER. This Lys-Asp-Glu-Leu (KDEL) tetrapeptide signal was found at the C terminus of several soluble proteins that are retained in the lumen of the ER. Moreover, removal of this sequence from a protein resulted in its secretion and the transfer of this signal to the C terminus of a normally secreted protein, lysozyme, resulted in its retention within the ER (MUNRO and PELHAM 1987). It should be noted however that recent experiments of ROSE and colleagues indicate that not all secretory proteins modified in such a manner remain sequestered in the ER, although their rate of transport from the ER to Golgi is significantly reduced (ZAGOURAS and ROSE 1989).

The argument for directed transport through the secretory pathway has been based at least in part on the observation that different proteins traverse the secretory pathway at different rates (LODISH et al. 1983). Thus it has been argued that slower proteins move at the bulk flow rate whereas faster proteins move at an accelerated rate because of efficient signal-dependent interactions with receptor molecules in the transport vesicles. ROTHMAN and colleagues (WIELAND et al. 1987) have attempted to measure the rate of bulk flow through the secretory pathway by measuring the apparent rate of secretion of small glycopeptides formed by incubating cells with *N*-acyl-Asn-[^{125}I]Tyr-Thr-NH_2. This cell-permeable peptide can act as an acceptor for *N*-linked glycosylation. Working on the assumption that such molecules would be glycosylated in the lumen of the ER and thereby rendered membrane impermeable, it was possible to estimate the time required for transport to the outside of the cell. The results of this study were interpreted as showing that the half-time for secretion of this bulk-phase marker was 10 min, as fast or faster than any known newly synthesized membrane protein or secreted protein. Thus it was concluded by the authors that with such a rapid and efficient bulk flow process in operation, there was no reason to invoke a requirement for secretory or plasma membrane targeting signals and that appropriate retention signals could explain the distribution of organellar constituents. KLAUSNER (1989) has pointed out, however, that two assumptions are required to accept the validity of this conclusion. The first is that glycosylation and release from cells of the tripeptide does in fact represent entrapment within the ER and transport as a fluid phase marker through the normal biosynthetic pathway. The second is that newly synthesized proteins enter the ER with a fully accessible and functional transport signal. The possibility that attaining a functional transport signal, due to conformational maturation, assembly, or posttranslational modification, might impose an obligatory lag period on this first step in the transport process is not considered. Clearly, for many proteins a considerable time is required for a stable, transport competent complex to form, and at least for some proteins this process of conformational maturation and oligomerization can wholly account for the lag in transport out of the ER (EINFELD and HUNTER 1988; GETHING et al. 1986).

An additional argument has been put forward to support the bulk flow/retention concept. This is that if protein transport from the ER to the Golgi and through the Golgi stacks does not require a signal, then the responsible transport vesicles should have no basis to select for and concentrate the transported molecules. Transported secretory proteins would be entrapped in the lumen of such vesicles at the same concentration at which they existed in the lumen of the ER or Golgi; transported integral membrane proteins would be at essentially the same density in the membranes of both the transport vesicles and their parental organelles. An analysis of transport vesicles budding from Golgi cisternae in vitro indicated that the concentration of a transported protein (the VSV G protein) was essentially the same in these vesicles as in the parental cisternae (ORCI et al. 1986), arguing against specific accumulation of the transported protein in the transport vesicle. It should be noted, however, that the VSV protein studied is present at unusually high concentrations within the Golgi and so this result may not be representative of the situation with other, less abundant proteins.

The bulk flow/retention model for transport of proteins through the secretory pathway is an attractive one because it provides a relatively simple mechanism for understanding how a broad array of disparate proteins can be transported through a complex vesicular system and how certain of these can be retained in specific organelles. Nevertheless, this model is hard to reconcile with a transport pathway possessing multiple branches, down which proteins are directed to organelles that are distinct from the plasma membrane. In addition, certain travelers of the secretory pathway possess distinct signals that specifically target them to discrete organelles. For example, the transport of certain hydrolases to the lysosome (and away from the secretory pathway) appears to require the addition of a phosphorylated carbohydrate moiety (Man-6-P CREEK and SLY 1984; HASILIK and NEUFELD 1980; SLY and FISCHER 1982), which in turn must require the recognition of signals within the polypeptide chain. Polypeptides whose secretion is regulated by specific stimuli are directed to and concentrated within secretory granules. The targeting of prosomatostatin is apparently controlled by the pro-peptide sequences that are removed within the secretory granule, since fusion of this region to globin results in its intracellular transport and sequestration within secretory vesicles (STOLLER and SHIELDS 1989).

In addition, in polarized epithelial cells different integral membrane proteins and secreted proteins are directed to distinct plasma membrane domains (apical and basolateral). Since proteins destined for different plasma membrane domains appear to travel through common vesicles to the trans-Golgi (FULLER et al. 1985), it seems likely that some form of transport signal must exist on these molecules to allow correct sorting to occur at that point. The location and identification of such signals are only now being determined (see COMPANS and SRINIVAS, this volume)

Thus, if the entire process of protein transport to the plasma membrane is considered, it seems likely that both bulk flow/retention and directed-transport mechanisms might be operating to target membrane and secretory proteins efficiently and specifically to their correct intracellular location. It is possible that bulk flow may operate on a subset of the system; only after proteins have entered the secretory pathway via a signal-dependent process and prior to directed transport from the trans-Golgi via vesicles that concentrate and target their protein cargo. Molecules lacking specific targeting or retention signals at this latter point in the pathway might be carried by default to the plasma membrane.

3.2 Transport Competency

Many proteins that are translocated into the lumen of the ER are not immediately transported on to the Golgi complex but reside in the ER while conformational or oligomeric maturity is achieved. It has been suggested, therefore, that proteins must become transport competent prior to exit from the ER. In some instances this may reflect the folding of the protein from the extended chain conformation, which thought to be translocated across the ER membrane, to the native conformation of the molecule. PCB proteins such as BiP may play important roles in this process, since, as we have discussed above, this protein appears to be capable of interacting with short peptide sequences similar to those that might be exposed on a newly translocated polypeptide chain. Although there is no direct genetic evidence for chaperonin-directed folding within the lumen of the ER, there is strong evidence that this type of process occurs within the mitochondria of yeast. OSTERMAN et al. (1989) have demonstrated in vitro that the mitochondrial hsp60 protein is necessary for the folding of dihydrofolate reductase (DHFR) following translocation into mitochondria and that this member of the heat shock family behaves as though it actively catalyzes the folding of this simple monomeric enzyme. In addition, CHENG et al. (1989) have described the isolation of a temperature-sensitive lethal mutant in yeast (*mif4*) that, at the nonpermissive temperature, correctly imported and proteolytically processed several mitochondrial enzymes which were then unable to assemble into functional oligomeric enzymes. Genetic mapping of the *mif4* locus has shown that this is identical to the *HSP60* gene and that a wild-type *HSP60* gene can complement the defect. Thus in this subcellular organelle a member of the chaperonin family is essential for a variety of oligomerization processes in vitro.

There is abundant evidence in mammalian cells pointing to the requirement for oligomer formation in order for a protein to attain transport competency. In lymphocytes of the B-cell lineage, for example, expression of immunoglobulin heavy chains in the absence of immunoglobulin light chains results in the heavy chains remaining in the ER. Without light chains, they are unable to be transported from the ER and a fraction of the molecules are found associated with BiP. Interestingly, fusion of such cells with cells producing light chains results in rapid dissociation from BiP and transport out of the ER (HENDERSHOT, personal communication), confirming the requirement for heavy and light chain association in development of transport competence. Similarly, the rate at which stable homo-oligomers of the Rous sarcoma virus (RSV) *env* gene product form is consistent with this being the determining factor in such molecules becoming transport competent and exiting the ER (EINFELD and HUNTER 1988). The molecular features of the monomeric polypeptides that cause them to be retained within the ER are not known. It is possible that the uncomplexed molecule exhibits sequences (analogous to the KDEL sequence) that specifically signal retention within this compartment. However, it seems more likely that in the absence of specific association the monomers nonspecifically aggregate with resident components in the ER. Indeed, mutant heavy chains that lack segments of the native untransported polypeptide can be transported out of the cell in the absence of light chains (HENDERSHOT et al. 1987), and truncations of the RSV *env* gene that result in removal of potential hydrophobic interactive domains result in efficient secretion of an apparently monomeric protein (EINFELD and HUNTER , in preparation).

Molecules that traverse the secretory pathway have presumably evolved to become transport competent following the formation of the species (monomer or multimer) that is biologically functional. Thus the various subunits of the acetylcholine receptor must be assembled into a functional hetero-oligomer before they can be transported to the plasma membrane. In contrast, expression of any one subunit alone results in it remaining in the ER (CLAUDIO et al. 1987; MISHINA et al. 1985). Thus if bulk flow/retention is operating to move both soluble and membrane-anchored proteins out of the ER, this suggests that there must be some mechanism for excluding from transport vesicles, or recycling to the ER, those molecules that are incompletely folded or incorrectly associated.

It has been suggested that *N*-linked glycosylation might play a role in conferring transport competence on some integral membrane proteins (GUAN et al. 1985; MACHAMER et al. 1985) and that this posttranslational modification might in some instances provide a signal for transport out of the ER. However, there are a growing number of examples of integral membrane proteins that are transported to the cell surface regardless of glycosylation (e.g., BARBOSA et al. 1987; HANNINK and DONOGHUE 1986) and it now seems more likely that the addition of an oligosaccharide chain may be required for the formation of the native conformation (and transport competency) of the protein. This conclusion is supported by the observation that in the case of the VSV G protein the stringent dependence on glycosylation for transport out of the ER can be suppressed by a mutation that introduces an acidic residue into the polypeptide chain (PITTA et al. 1989).

It seems likely, then, that many processes that ultimately impact on protein conformation (glycosylation, protein folding, and oligomer formation) can influence the transport competence of any particular molecule. In some instances, mutations that affect the oligomerization of proteins, such as that observed with mutant HA of influenza virus, might be expected to interfere with transport (GETHING et al. 1986), but in other instances, mutations which are equally effective at blocking transit from the ER induce alterations in the protein that are much more subtle and more difficult to define (DOMS et al. 1988). The molecular definition of the interactions and ER components that mediate the efficient retention of proteins in this compartment remains an important area for understanding this critical stage of protein transport.

3.3 Protein Domains that Influence Interactions with the Transport Pathway

3.3.1 The Signal Peptide

Different regions of a protein may influence the manner and efficiency with which it traverses the secretory pathway, and to some extent this differs between soluble, secreted proteins and those that span the membrane. The initial entry of both types of protein into the secretory pathway, however, required the interaction of a signal peptide sequence on the polypeptide with recognition and translocation machinery. The signal peptide is in most cases a transient extension at the N terminus of the protein and is removed by a member of a class of enzymes known as signal peptidases once its targeting function has been carried out.

Signal peptides have three distinct domains (Fig. 3): and N-terminal positively charged region (the n-region, 1–5 residues long), a central, hydrophobic section (the h-region, 7–15 residues long), and a more polar C-terminal domain (c-region, 3–7 residues long). There is no precise sequence conservation beyond this overall pattern. The presence of a positively charged lysine or arginine in the n-region is universal among bacterial signal peptides and is found in most but not all eukaryotic sequences. It has been suggested by VON HEIJNE (1984) that the non-formylated, charged amino group on the initiating

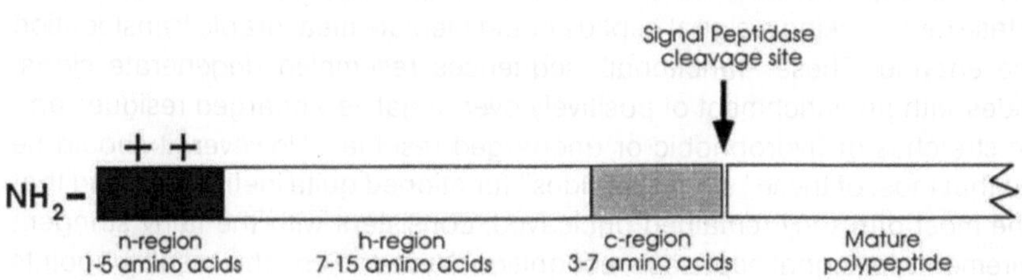

Fig. 3. Schematic representation of the signal peptide functional regions, adapted from von HEIJNE (1986). In some membrane proteins, for example the *env* gene product of several retroviruses, the n-region of the signal peptide can be longer than 30 amino acids (HUNTER 1988)

methionine might substitute for the positively charged amino acids in higher organisms. In bacterial proteins, the removal of the positively charged N terminus has been shown to slow export, but no absolute requirement has been demonstrated (BOSCH et al. 1989; PUZISS et al. 1989). In yeast invertase and prepro-α factor, substitution of other amino acids for basic residues within the n-region had no effect on secretion (BROWN et al. 1984; GREEN et al. 1989), supporting the conclusions of VON HEIJNE (1984) from sequence analyses. The h-regin, on the other hand, appears to be critical for translocation of the polypeptide into the lumen of the RER. While most of the evidence for this functional role has come from studies in *E. coli*, where mutations that disrupt the h-region by introduction of charged or helix-breaking residues result in defective translocation (for review see SILHAVY et al. 1983), similar studies in yeast have yielded similar results (ALLISON and YOUNG 1988; ALLISON and YOUNG 1989; KAISER and BOTSTEIN 1986). Similarly, deletion mutations that reduce the length of the h-region of the RSV envelope glycoprotein and the bovine parathyroid hormone signal peptides result in proteins that are unable to be translocated (CLIOFI et al. 1989; HUNTER 1988). The length and α-helical propensity of this region thus appear to be important for its function, possibly because it is this portion of the signal peptide that must span the hydrophobic environment of the lipid bilayer.

The major purpose of the c-region sequence appears to be in defining the site of signal peptidase cleavage, since modifications that block cleavage have no effect on the translocation process itself (NOTHWEHR et al. 1989; WILLS et al. 1983). PERLMAN and HALVORSON (1983) and VON HEIJNE (1983) have examined sequences of a number of membrane proteins and have described amino acid sequence patterns that allow prediction of signal peptidase cleavage sites with > 90% accuracy. The most striking feature of signal peptidase cleavage sites is the presence of an amino acid with a small, uncharged side chain at the C terminus of the signal peptide. The most common amino acids found at this position are alanine and glycine. From the statistical analyses noted above and from mutational analyses (NOTHWEHR and GORDON 1989; WILLS et al. 1983), the peptidase cleavage site appears to be determined by sequences within the signal peptide and not by sequences beyond the cleavage site.

That signal peptides are highly variable structures with only a minimal amount of sequence conservation is reinforced by the finding of KAISER et al. (1987) that approximately 20% of random sequences cloned in front of an invertase gene lacking a signal peptide could mediate measurable translocation of the enzyme. These "functional" sequences resembled degenerate signal peptides with an enrichment of positively over negatively charged residues and some stretches of hydrophobic or uncharged residues. However, it should be noted that most of these "signal peptides" functioned quite inefficiently and that for the most part they remained uncleaved, consistent with the fairly stringent requirements for signal peptidase recognition. Nevertheless, this result pinpoints the problems of identifying *cis*-acting signals within a polypeptide chain that can play important roles in the intracellular transport process.

3.3.2 Stop Translocation/Anchor Sequences

The feature that distinguishes integral membrane proteins from those that are secreted from a eukaryotic cell is the presence of a membrane-spanning anchor domain. This domain is not apparently required for the acquisition of transport competency, since its removal usually results in conversion of the membrane-bound protein into a secreted molecule (reviewed in HUNTER 1988). Nevertheless, it can influence the efficiency with which the protein is transported through the cell and its final destination within the cell.

Integral membrane proteins exist in a variety of transmembrane topologies (Fig. 4). Proteins that cross the membrane a single time have been termed bitopic, and those with two or more membrane-spanning domains polytopic. The former can be subdivided into three classes (I-III) based on their mode of insertion and their orientation within the membrane. Class I proteins are synthesized with a cleavable N-terminal signal peptide and a stop transfer/membrane anchor sequence (typically a stretch of 18–30 hydrophobic amino acids followed by one or more positively charged residues) located towards the C terminus of the polypetide chain. The final orientation of this class of proteins, following removal of the signal peptide, is with the N terminus outside the cell and the C terminus in the cytoplasm (N_{out}-C_{in}). Several integral membrane proteins remain anchored to the membrane by a long uncleaved signal peptide that may be located at any point within the protein. In class II proteins this results in a molecule with a C_{out}-N_{in} topology, while class III integral membrane proteins, also anchored by an uncleaved signal peptide, have the opposite orientation (N_{out}-C_{it}). This latter class of proteins have an inverted charge polarity across their hydrophobic region compared to most signal peptides, with positively charged residues clustered at its C-terminal side. Recent experiments have demonstrated a strong correlation between the transmembrane topology of

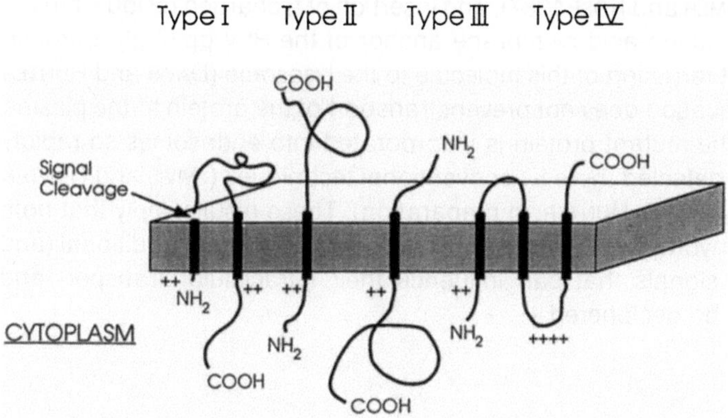

Fig. 4. Topological organization of different types of integral membrane proteins. In some type IV proteins which span the membrane multiple times, the first signal peptide may be removed by signal peptidase. (adapted from VON HEIJNE (1986))

integral membrane proteins and the distribution of positively charged residues around the hydrophobic domain, positively charged domains being found predominantly on the inner side of the membrane (VON HEIJNE 1986). LODISH and coworkers (HARTMANN et al. 1989) have suggested that the charge distribution around the first hydrophobic, transmembrane domain may be critical for establishing the topology of eukaryotic membrane proteins, including the type IV polytopic molecules, and the results of experimental manipulations of the charge cluster location are consistent with this conclusion (HAEUPTLE et al. 1989; SZCZESNA-SKOPURA and KEMPER 1989).

In contrast to the integral membrane proteins that are anchored by a stretch of hydrophobic amino acids, a class of proteins has been described in which the external polypeptide domain is anchored in the plasma membrane by a glycosylphosphatidylinositol (GPI) moiety. These GPI-anchored proteins are synthesized with hydrophobic C-terminal domains that are removed almost immediately after translation. A preformed GPI molecule is then covalently linked to the new C-terminal amino acid. Experiments where the C-terminal sequences from a GPI-anchored protein have been substituted for the hydrophobic amino acid anchor sequence of type I proteins, or have been added to a soluble protein, have shown that the hybrid protein can be efficiently linked to GPI and transported to the cell surface (CARAS and WEDDELL 1989; CARAS et al. 1989; CRISE et al. 1989). However, at least in the case of VSV G protein, the GPI-modified protein is then transported to the apical rather than the basolateral membrane of polarized epithelial cells (BROWN et al. 1989).

Thus the mechanism of membrane anchoring can determine the sorting of proteins to the apical or basolateral surface of a polarized cell. In addition, the membrane anchor domain of proteins can play other roles in determining the final intracellular location of a protein. As discussed previously, removal of the first transmembrane domain of the polytopic infectious bronchitis virus E1 protein results in its transport to the plasma membrane rather than retention in the Golgi (MACHAMER and ROSE 1987), and insertion of a charged residue into the middle of the 27-amino-acid membrane anchor of the RSV gp37 glycoprotein results in efficient targeting of this molecule to the lysosome (DAVIS and HUNTER 1987). This modification does not prevent transport of the protein to the plasma membrane, but the mutant protein is incorporated into endosomes so rapidly that it cannot be detected there by conventional techniques (DAVIS and HUNTER 1987; P. JOHNSON and E. HUNTER, in preparation). These results imply that both protein and lipid hydrophobic transmembrane domains contain additional (and perhaps subtle) signals that can influence their intracellular transport and which remain to be deciphered.

3.3.3 Cytoplasmic Domains

A protein-anchored integral membrane proteins generally consists of three domains: the ectodomain, the transmembrane anchor domain, and the cytoplasmic domain. Because of the position of the latter it was initially considered a good candidate for directing or mediating the targeting of transport vesicles within the secretory pathway. It has therefore been the target of several mutagenesis approaches in a variety of different systems. These studies have yielded a complicated pattern of results from which it is clear that the cytoplasmic domain can be important in facilitating transport of some but not all integral membrane proteins. As we have discussed previously, PAABO et al. (1987) identified an ER retention sequence within the cytoplasmic domain of the adenovirus E19 glycoprotein. Deletion of 9 amino acids from the 15-amino-acid domain resulted in expression of this normally ER-localized protein on the cell surface (PAABO et al. 1987), and transfer of the C-terminal 6 residues to a second protein resulted in its retention in the ER (NILSSON et al. 1989).

In contrast, in studies of the VSV G protein, ROSE and coworkes showed that deleting residues from the 29-amino-acid cytoplasmic domain caused large delays in the normally rapid transport of the protein from the ER to the Golgi and that substitution of 12 foreign amino acids for the normal cytoplasmic sequence of this protein resulted in absolute retention within the ER (ROSE and BERGMANN 1983). Different results have been obtained in other systems. In the case of the HA of influenza virus and the gp37 of RSV, mutations that led to a truncation of the cytoplasmic domain and substitution of other amino acids resulted in a reduction in the transport rate from the ER (DOYLE et al. 1985; WILLS et al. 1984). On the other hand, complete deletions of the cytoplasmic domain of these two proteins had little effect on the transport of these glycoproteins (DOYLE et al. 1986; PEREZ et al. 1987), suggesting that the inhibitory effects observed earlier resulted from the amino acid substitutions rather than the loss of the cytoplasmic domain itself.

In some instances mutations within the cytoplasmic domain can have profound effects on the extracellular domain of the protein. GETHING et al. (1986) demonstrated that mutations within the cytoplasmic domain of the influenza virus HA could affect the conformation of the extracellular domain by preventing assembly and trimerization of the HA molecule. This correlated well with the failure of those mutants to be efficiently transported and provided an explanation for the phenotype. However, in other situations the deleterious effect of cytoplasmic domain mutations is less obvious. DOMS et al. (1988) analyzed five VSV G proteins with a variety of cytoplasmic mutations that resulted in range of phenotypes from slow transport to a complete block in transport from the ER. Three assays were used to examine glycoprotein folding and trimerization of these mutants, and it was found that the ectodomains were able to fold and trimerize correctly. Even in mutants in which transport from the ER was completely blocked or in which exit from the ER took 8 times longer than in the case of the wild-type protein, folding and trimerization occurred at least as rapidly as in the wild-type protein. Rose and colleagues (DOMS et al. 1988; ROSE and DOMS 1988) have concluded that with the VSV G protein the cytoplasmic domain acts to facilitate transport out of the ER, and that certain modifications might result in retention signals similar to that found in the adenovirus E19 protein. For some proteins, then, the cytoplasmic domain may play a relatively passive role in intracellular transport while being functionally important for the protein itself, and for others the cytoplasmic domain may facilitate the transport process. For different retroviral glycoproteins this spectrum of effects can be observed. Both the RSV and HIV *env* gene products can be efficiently transported to the cell surface with truncated cytoplasmic domains, even though the latter protein, which normally has a long (> 150 amino acids) cytoplasmic domain, is then no longer able to confer infectivity on the virus (J. Dubay and E. Hunter unpublished). Complete truncation of the cytoplasmic domain of the gp22 (TM) of Mason-Pfizer monkey virus, on the other hand, blocks its normal transport from the ER (B. Brody and E. Hunter unpublished). It is thus likely that different interactions are involved for each protein under study, making the process of dissecting out the critical components more difficult.

4 Concluding Remarks

We have attempted to provide an overview of the cellular pathway that proteins must traverse in order to be presented on (or secreted from) the plasma membrane of a cell. In addition, we have addressed the structural aspects of the transported molecules themselves that might interact with this pathway and influence its efficiency. *In vitro* studies of reconstituted systems are clearly providing new insight into the molecular processes that allow vesicle formation and fusion within the secretory pathway. Nevertheless, the identification of the molecular mechanisms involved in targeting nascent vesicles to the next stage of the pathway or to a specific location within the cell remains a critical problem.

As in most biological systems the acquisition of knowledge about the pathway for transporting proteins to the cell surface has established a new appreciation of the complexity of the system under study.

Despite being recognized for over 20 years as a component of a protein that is vital for directing it into the secretory pathway, the signal peptide remains somewhat of an enigma. Its complex functional role, coupled with cross-linking information, indicates that it must interact with proteins, but whether it mediates transfer of the nascent polypeptide chain through a protein pore or directly through the lipid bilayer remains to be determined.

The role of protein ternary structure and transport competence has been discussed above, but how the secretory pathway senses when a protein is competent for transport and what ER proteins are involved in this crucial retention event is not known. These unresolved problems impinge on attempts to determine the relative roles of bulk flow/retention and signal-directed transport within the pathway. How are proteins prevented from entering transport vesicles if they are to be retained within an organelle; what are the additional signals that are required, and where in the protein are they located? Similarly, what signals direct proteins to various final destinations after the secretory pathway branches following the trans-Golgi complex?

Much still remains to be understood about these finer points of plasma membrane transport, and it is likely that more refined reconstituted systems coupled with genetic probes will allow answers to many of these questions. Progress in understanding the simpler and more readily definable secretory pathway of yeast may also point to new avenues for research in higher eukaryotes.

References

Ainger K, Meyer DI (1986) Translocation of nascent secretory proteins across membranes can occur late in translation. EMBO J 5: 951–955

Allison DS, Young ET (1988) Single-amino-acid substitutions within the signal sequence of yeast prepro-alpha-factor affect membrane translocation. Mol Cell Biol 8: 1915–1922

Allison DS, Young ET (1989) Mutations in the signal sequence of prepro-alpha-factor inhibit both translocation into the endoplasmic reticulum and processing by signal peptidase in yeast cells. Mol Cell Biol 9: 4977–4985

Baker RK, Lively MO (1987) Purification and characterization of hen oviduct microsomal signal peptidase. Biochemistry 26: 8561–8567

Balch WE, Elliot MM, Keller DS (1986) ATP-coupled transport of vesicular stomatitis virus G protein between the endoplasmic reticulum and the Golgi. J Biol Chem 261: 14681–14689

Barbosa JA, Santos AJ, Mentzer SJ, Strominger JL, Burakoff SJ, Biro PA (1987) Site-directed mutagenesis of class I HLA genes. Role of glycosylation in surface expression and functional recognition. J Exp Med 166: 1329–1350

Beckers CJM, Balch WE (1989) Calcium and GTP: essential components in vesicular trafficking between the endoplasmic reticulum and Golgi apparatus. J Cell Biol 108: 1245–1256

Beckers CJM, Keller DS, Balch WE (1987) Semi-intact cells permeable to macromolecules: use in reconstitution of protein transport from the endoplasmic reticulum to the Golgi complex. Cell 50: 523–534

Beckers CJM, Block MR, Glick BS, Rothman JE, Balch WE (1989) Vesicular transport between the endoplasmic reticulum and the Golgi stack requires the NEM-sensitive fusion protein. Nature 339: 397–398

Bennett MK, Wandinger NA, Simons K (1988) Release of putative exocytic transport vesicles from perforated MDCK cells. EMBO J 7: 4075–4085

Berger M, Schmidt MF(1985) Protein fatty acyltransferase is located in the rough endoplasmic reticulum. FEBS Lett 187: 289–294

Block MR, Glick BS, Wilcox CA, Wieland FT, Rothman JE (1988) Purification of an N-ethylmaleimide-sensitive protein catalyzing vesicular transport. Proc Natl Acad Sci USA 85: 7852–7856

Boehni PC, Deshaies RJ, Schekman R (1988) SEC11 is required for signal peptide processing and yeast cell growth. J Cell Biol 106: 1035–1042

Bole DG, Hendershot LM, Kearney JF (1986) Post-translational association of immunoglobulin heavy chain binding protein with nascent heavy chains in nonsecreting and secreting hybridomas. J Cell Biol 102: 1558–1566

Bosch D, de Boer P, Bitter W, Tommassen J (1989) The role of the positively charged N-terminus of the signal sequence of *E. coli* outer membrane protein PhoE in export. Biochim Biophys Acta 979: 69–76

Brown DA, Crise B, Rose JK (1989) Mechanism of membrane anchoring affects polarized expression of two proteins in MDCK cells. Science 245: 1499–1501

Brown PA, Halvorson HO, Raney P, Perlman D (1984) Conformational alterations in the proximal portion of the yeast invertase signal peptide do not block secretion. Mol Gen Genet 197: 351–357

Bulleid NJ, Freedman RB (1988) Defective co-translational formation of disulphide bonds in protein disulphide-isomerase-deficient microsomes. Nature 335: 649–651

Caras IW, Weddell GN (1989) Signal peptide for protein secretion directing glycophospholipid membrane anchor attachment. Science 243: 1196–1198

Caras IW, Weddell GN, Williams SR (1989) Analysis of the signal for attachment of a glycophospholipid membrane anchor. J Cell Biol 108: 1387–1396

Chao CCK, Bird P, Gething MJ, Sambrook J (1987) Post-translational translocation of influenza virus hemagglutinin across microsomal membranes. Mol Cell Biol 7: 3842–3845

Cheng MY, Hartl F-U, Martin J, Pollock, RA, Kalousek F, Neupert W, Hallberg EM, Hallberg RL, Horwich AL (1989) Mitochondrial heat-shock protein hsp60 is essential for assembly of proteins imported into yeast mitochondria. Nature 337: 620–625

Chirico WJ, Waters MG, Blobel G (1988) 70K heat shock related proteins stimulate protein translocation into microsomes. Nature 332: 805–810

Chung KN, Walter P, Aponte GW, Moore HP (1989) Molecular sorting in the secretory pathway. Science 243: 192–197

Cioffi J A, Allen KL, Lively MO, Kemper B (1989) Parallel effects of signal peptide hydrophobic core modifications on co-translational translocation and post-translational cleavage by purified signal peptidase. J Biol Chem 264: 15052–15058

Claudio T, Green WN, Hartman DS, Hayden D, Paulson HL (1987) Genetic reconstitution of functional acetylcholine receptor channels in mouse fibroblasts. Science 238: 1688–1694

Connolly T, Gilmore R (1986) Formation of a functional ribosome-membrane junction during translocation requires the participation of a GTP-binding protein. J Cell Biol 103: 2253–2261

Connolly T, Gilmore R (1989) The signal recognition particle receptor mediates the GTP-dependent displacement of SRP from the signal sequence of the nascent polypeptide. Cell 57: 599–610

Copeland CS, Doms RW, Bolzau EM, Webster RG, Helenius A (1987) Assembly of influenza hemagglutinin trimers and its role in intracellular transport. J Cell Biol 103: 1179–1191

Copeland CS, Zimmer K-P, Wagner KR, Healey GA, Mellman I, Helenius A (1988) Folding, trimerization, and transport are sequential events in the biogenesis of influenza virus hemagglutinin. Cell 53: 197–209

Creek KE, Sly WS (1984) Lysosomes in pathology and biology. In: Dingle JT, Dean RT, Sly W (ed) Lysosomes in pathology and biology. Elsevier/North Holland, New York, pp 63–82

Crise B, Ruusala A, Zagouras P, Shaw A, Rose JK (1989) Oligomerization of glycolipid-anchored and soluble forms of the vesicular stomatitis virus glycoprotein. J Virol 63: 5328–5333

Date T, Wickner W (1981) Isolation of the *Escherichia coli* leader peptidase gene and effects of leader peptidase overproduction in vivo. Proc Natl Acad Sci USA 78: 6106–6110

Davis GL, Hunter E (1987) A charged amino acid substitution within the transmembrane anchor of the Rous sarcoma virus envelope glycoprotein affects surface expression but not intracellular transport. J Cell Biol 105: 1191–1203

Deshaies RJ, Schekman R (1987) A yeast mutant defective at an early stage in import of secretory protein precursors into the endoplasmic reticulum. J Cell Biol 105: 633–645

Deshaies RJ, Koch BD, Werner-Washburne M, Craig EA, Schekman R (1988) A subfamily of stress proteins facilitates translocation of secretory and mitochondrial precursor polypeptides. Nature 332: 800–805

Doms RW, Ruusala A, Machamer C, Helenius J, Helenius A, Rose JK (1988) Differential effects of mutations in three domains on folding, quaternary structure, and intracellular transport of vesicular stomatitis virus G protein. J Cell Biol 107: 89–99

Doms RW, Russ G, Yewdell JW (1989) Brefeldin A redistributes resident and itinerant Golgi proteins to the endoplasmic reticulum. J Cell Biol 109: 61–72

Doyle C, Roth MG, Sambrook J, Gething M-J (1985) Mutations in the cytoplasmic domain of influenza virus hemagglutinin affect different stages of intracellular transport. J Cell Biol 100: 704–714

Doyle C, Sambrook J, Gething M-J (1986) Analysis of progressive deletions of the transmembrane and cytoplasmic domins of influenza hemagglutinin. J Cell Biol 103: 1193–1204

Dunphy WG, Rothman JE (1985) Compartmental organization of the Golgi stack. Cell 42: 13–21

Einfeld D, Hunter E (1988) Oligomeric structure of a prototype retrovirus glycoprotein. Proc Natl Acad Sci USA 85: 8688–8692

Ellis RJ (1987) Proteins as molecular chaperones. Nature 328: 378–379

Ellis RJ, van der Vies SM, Hemingsen SM (1989) The molecular chaperone concept. Biochem Soc Symp 55: 145–153

Evans EA, Gilmore R, Blobel G (1986) Purification of microsomal signal peptidase as a complex. Proc Natl Acad Sci USA 83: 581–585

Ferguson MAJ, Williams AF (1989) Cell-surface anchoring of proteins via glycosyl-phosphatidylinositol structures. Annu Rev Biochem 57: 285–320

Fitting T, Kabat D (1982) Evidence for a glycoprotein 'signal' involved in transport between subcellular organelles. J Biol Chem 257: 14011–14017

Freedman RB (1989) Protein disulfide isomerase: multiple roles in the modification of nascent secretory proteins. Cell 57: 1069–1072

Fujiwara T, Oda K, Yokota S, Takatsuki A, Ikehara Y (1988) Brefeldin A causes disassembly of the Golgi complex and accumulation of secretory proteins in the endoplasmic reticulum. J Biol Chem 263: 18545–18552

Fuller SD, Bravo R, Simons K (1985) An enzymatic assay reveals that proteins destined for the apical or basolateral domains of an epithelial cell line share the same late Golgi compartments. EMBO J 4: 297–307

Geetha-Habib M, Noiva R, Kaplan HA, Lennarz WJ (1988) Glycosylation site binding protein, a component of oligosaccharyl transferase, is highly similar to three other 57 kd luminal proteins of the ER. Cell 54: 1053–1060

Gething M-J, McCammon K, Sambrook J (1986) Expression of wild-type and mutant forms of influenza hemagglutinin: the role of folding in intracellular transport. Cell 46: 939–950

Gilmore R, Blobel G (1983) Transient involvement of signal recognition particle and its receptor in the microsomal membrane prior to protein translocation. Cell 35: 677–685

Gilmore R, Blobel G, Walter P (1982) Protein translocation across the endoplasmic reticulum. I. Detection in the microsomal membrane of a receptor for the signal recognition particle. J Cell Biol 95: 463–469

Green R, Kramer RA, Shields D (1989) Misplacement of the amino-terminal positive charge in the prepro-alpha-factor signal peptide disrupts membrane translocation in vivo. J Biol Chem 264: 2963–2968

Griffiths G, Simons K (1986) The trans Golgi network: sorting at the exit site of the Golgi complex. Science 234: 438–443

Griffiths G, Pfeiffer S, Simons K, Matlin K (1985) Exit of newly synthesized membrane proteins from the trans cisterna of the Golgi complex to the plasma membrane. J Cell Biol 101: 949–964

Guan J-L, Machamer CE, Rose JK (1985) Glycosylation allows cell-surface transport of an anchored secretory protein. Cell 42: 489–496

Gumbiner G, Kelly RB (1982) Two distinct intracellular pathways transport secretory and membrane glycoproteins to the surface of pituitary tumor cells. Cell 28: 51–59

Haeuptle MT, Flint N, Gough NM, Dobberstein B (1989) A tripartite structure of the signals that determine protein insertion into the endoplasmic reticulum membrane. J Cell Biol 108: 1227–1236

Hannink M, Donoghue DJ (1986) Cell surface expression of membrane-anchored v-sis gene products: glycosylation is not required for cell surface transport. J Cell Biol 103: 2311–2322

Hartmann E, Rapoport TA, Lodish HF (1989) Predicting the orientation of eukaryotic membrane-spanning proteins. Proc Natl Acad Sci USA 86: 5786–5790

Hasilik A, Neufeld EF (1980) Biosynthesis of lysosomal enzymes in fibroblasts: phosphorylation of mannose residues. J Biol Chem 255: 4946–4950

Hendershot L, Bole D, Koehler G, Kearney JF (1987) Assembly and secretion of heavy chains that do not associate posttranslationally with immunoglobulin heavy chain-binding protein. J Cell Biol 104: 761–767

Hortsch M, Avossa D, Meyer DI (1985) A structural and functional analysis of the docking protein. J Biol Chem 260: 9137–9145

Hortsch M, Avossa D, Meyer DI (1986) Characterization of secretory protein translocation: ribosome-membrane interaction in endoplasmic reticulum. J Cell Biol 103: 241–253

Hunter E (1988) Membrane insertion and transport of viral glycoproteins: a mutational analysis. In: Robbins PW (ed) Protein transfer and organelle biogenesis. Das RC, Academic, pp 109-158

Jackson RC, White WR (1981) Phospholipid is required for the processing of presecretory proteins by detergent-solubilized canine pancreatic signal peptidase. J Biol Chem 256: 2545–2550

Jing S, Trowbridge IS (1987) Identification of the intermolecular disulfide bonds of the human transferrin receptor and its lipid-attachment site. EMBO J 6: 327–331

Kaiser CA, Botstein D (1986) Secretion-defective mutations in the signal sequence for *Saccharomyces cerevisiae* invertase. Mol Cell Biol 6: 2382–2391

Kaser CA, Preuss D, Grisafi P, Botstein D (1987) Many random sequences functionally replace the secretion signal sequence of yeast invertase. Science 235: 312–317

Kaufman JF, Krangel MS, Strominger JL (1984) Cysteines in the transmembrane region of major histocompatibility complex antigens are fatty acylated via thioester bonds. J Biol Chem 259: 7230–7238

Kelly RB (1985) Pathway of protein secretion in eukaryotes. Science 230: 139–145

Klausner RD (1989) Sorting and traffic in the central vacuolar system. Cell 57: 703–706

Kornfeld R, Kornfeld S (1985) Assembly of asparagine-linked oligosaccharides. Annu Rev Biochem 54: 631–644
Kozutsumi Y, Segal M, Normington K, Gething M-J, Sambrook J (1988) The presence of malfolded proteins in the endoplasmic reticulum signals induction of glucose-regulated proteins. Nature 332: 462–464
Krieg UC, Walter P, Kohnson AE (1986) Photocrosslinking of the signal sequence of nascent preprolactin to the 54-kilodalton polypeptide of the signal recognition particle. Proc Natl Acad Sci USA 83: 8604–8608
Kurzchalia TV, Wiedmann M, Girshovich AS, Bochkareva ES, Bielka H, Rapoport TA (1986) The signal sequence of nascent preprolactin interacts with the 54K polypeptide of the signal recognition particle. Nature 320: 634–636
Lecker S, Lill R, Ziegelhoffer T, Georgopoulos C, Bassford P, Kumamoto CA, Wickner W (1989) Three pure chaperone proteins of *Escherichia coli*—SecB, trigger factor, GroEl—form soluble complexes with precursor protein in vitro. EMBO J 8: 2703–2709
Ledford BE, Davis DF (1983) Kinetics of serum protein secretion by hepatoma cells: evidence for multiple secretory pathways. J Biol Chem 258: 3304–3308
Lippincott-Schwartz J, Yuan LC, Bonifacino JS, Klausner RD (1989) Rapid redistribution of Golgi proteins into the ER in cells treated with brefeldin A: evidence for membrane cycling from Golgi to ER. Cell 56: 801–813
Lively MO, Walsh KA (1983) Hen oviduct signal peptidase is an integral membrane protein. J Biol Chem 258: 9488–9495
Lodish HF, Kong N, Snider M, Strous GJAM (1983) Hepatoma secretory proteins migrate from rough endoplasmic reticulum to Golgi at characteristic rates. Nature 304: 80–83
Lodish HF, Kong N, Hirani S, Rasmussen J (1987) A vesicular intermediate in the transport of hepatoma secretory proteins from the rough endoplasmic reticulum to the Golgi complex. J Cell Biol 104: 221–230
Low MG, Saltiel AR (1988) Structural and functional roles of glycosyl-phosphatidylinositol in membranes. Science 239: 268–275
Machamer CE, Rose JK (1987) A specific transmembrane domain of a coronavirus E1 glycoprotein is required for its retention in the Golgi region. J Cell Biol 105: 1205–1214
Machamer CE, Florkiewicz RZ, Rose JK (1985) A single N-linked oligosaccharide at either of the two normal sites is sufficient for transport of vesicular stomatitis virus G protein to the cell surface. Mol Cell Biol 11: 3074–3083
Mack D, Berger M, Schmidt MT, Kruppa J (1987) Cell free fatty acylation of microsomal integrated and detergent-solubilized glycoprotein of vesicular stomatitis virus. J Biol Chem 262: 4297–4302
Malhotra V, Orci L, Glick MR, Rothman JE (1988) Role of an N-ethylmaleimide sensitive transport component for promoting fusion of transport vesicles with cisternae of the Golgi stack. Cell 54: 221–227
Matlin KS, Simons K (1983) Reduced temperature prevents transfer of a membrane glycoprotein to the cell surface but does not prevent terminal glycosylation. Cell 34: 233–243
Mayer T, Tamura T, Falk M, Niemann H (1988) Membrane insertion and intracellular transport of the coronavirus glycoprotein, E1, a class III membrane glycoprotein. J Biol Chem 263: 14956–14963
Melancon P, Glick BS, Malhotra V, Wiedman PJ, Sarafini T, Gleason ML, Orci L, Rothman JE (1987) Involvement of GTP-binding "G" proteins in transport through the Golgi stack. Cell 51: 1053–1062
Meyer DI, Krause E, Dobberstein B (1982) Secretory protein translocation across membranes—the role of docking protein. Nature 297: 647–650
Mishina M, Tobimatsu T, Imoto K, Tanaka K, Fujita Y (1985) Location of functional regions of acetylcholine receptor alpha-subunit by site directed mutagenesis. Nature 313: 364–369
Misumi Y, Misumi Y, Miki K, Takatsuki A, Tamura G, Ikehara Y (1986) Novel blockade by brefeldin A of intracellular transport of secretory proteins in cultured rat hepatocytes. J Biol Chem 261: 11398–11403
Moore H-PH, Kelly RB (1986) Re-routing of a secretory protein by fusion with human growth hormone sequence. Nature 321: 443–446
Mueckler M, Lodish HF (1986) The human glucose transporter can insert post-translationally into microsomes. Cell 44: 629–637
Munro S, Pelham HRB (1986) An Hsp 70-like protein in the ER: identity with the 78 kd glucose-regulated protein and immunoglobulin heavy chain binding protein. Cell 46: 291–300
Munro S, Pelham HRB (1987) A C-terminal signal prevents secretion of luminal ER proteins. Cell 48: 899–907

Nicchitta CV, Blobel G (1989) Nascent secretory chain binding and translocation are distinct processes: differentiation by chemical alkylation. J Cell Biol 108: 789–795

Nilsson T, Jackson M, Peterson PA (1989) Short cytoplasmic sequences serve as retention signals for transmembrane proteins in the endoplasmic reticulum. Cell 58: 707–718

Normington K, Kohno K, Kozutisumi Y, Gething M-j, Sambrook J (1989) *Saccharomyces cerevisiae* encodes an essential protein homologous in sequence and function to mammalian BiP. Cell 57: 1223–1236

Nothwehr SF, Gordon JI (1989) Eukaryotic signal peptide structure/function relationships. Identification of conformational features which influence the site and efficiency of cotranslational proteolytic processing by site-directed mutagenesis of human pre(Δ pro)apolipoprotein A-II. J Biol Chem 264: 3979–3987

Nothwehr SF, Folz RJ, Gordon JI (1989) Uncoupling of co-translational translocation from signal peptidase processing in a mutant rat preapolipoprotein-A-IV with a deletion that includes the COOH-terminal region of its signal peptide. J Biol Chem 264: 4642–4647

Orci L, Ravazzola M, Amherdt M, Madsen O, Vassalli J-D, Perrelet A (1985) Direct identification of prohormone conversion site in insulin-secreting cells. Cell 42: 671–681

Orci L, Glick BS, Rothman JE (1986) A new type of coated vesicular carrier that appears not to contain clathrin: its possible role in protein transport within the Golgi stack. Cell 46: 171–184

Orci L, Ravazzola M, Amherdt M, Perrelet A, Powell SK, Quinn DL, Moore H-PH (1987) The trans-most cisternae of the Golgi complex: a compartment for sorting of secretory and plasma membrane proteins. Cell 51: 1039–1051

Orci L, Malhotra V, Amherdt M, Serafini T, Rothman JE (1989) Dissection of a single round of vesicular transport: sequential intermediates for intercisternal movement in the Golgi stack. Cell 56: 357–368

Ostermann J, Horwich AL, Neupert W, Hartl F-U (1989) Protein folding in mitochondria requires complex formation with hsp60 and ATP hydrolysis. Nature 341: 125–130

Paabo S, Bhat BM, Wold WS, Peterson PA (1987) A short sequence in the COOH-terminus makes an adenovirus membrane glycoprotein a resident of the endoplasmic reticulum. Cell 50: 311–317

Palade G (1975) Intracellular aspects of the process of protein synthesis. Science 189: 347–358

Perara E, Rothman RE, Lingappa VR (1986) Uncoupling translocation from translation: implications for transport of proteins across membranes. Science 232: 348–352

Perez LG, Davis GL, Hunter E (1987) Mutants of the Rous sarcoma virus envelope glycoprotein that lack the transmembrane anchor and/or cytoplasmic domains: analysis of intracellular transport and assembly into virions. J Virol 61: 2981–2988

Perlman D, Halvorson HO (1983) A putative signal peptidase recognition site and sequence in eucaryotic and procaryotic signal peptides. J Mol Biol 167: 391–409

Pitta AM, Rose JK, Machamer CE (1989) A single-amino-acid substitution eliminates the stingent carbohydrate requirement for intracellular transport of a viral glycoprotein. J Virol 63: 3801–3809

Poruchynsky MS, Tyndall C, Both GW, Sato F, Bellamy AR, Atkinson PH (1985) Deletions into an NH_2-terminal hydrophobic domain result in secretion of rotavirus VP7, a resident endoplasmic reticulum glycoprotein. J Cell Biol 101: 2199–2209

Puziss JW, Fikes JD, Bassford PJ (1989) Analysis of mutational alterations in the hydrophilic segment of the maltose-binding protein signal peptide. J Bacteriol 171: 2303–2311

Reading DS, Hallberg RL, Myers AM (1989) Characterization of the yeast HSP60 gene coding for a mitochondrial assembly factor. Nature 337: 655–659

Robinson A, Kaderbhai MA, Austen BM (1987) Identification of signal sequence binding proteins integrated into the rough endoplasmic reticulum membrane. Biochem J 242: 767–777

Rose JK, Bergmann JE (1983) Altered cytoplasmic domains affect intracellular transport of the vesicular stomatitis virus glycoprotein. Cell 30: 513–524

Rose JK, Doms RW (1988) Regulation of protein export from the endoplasmic reticulum. Annu Rev Cell Biol 4: 257–288

Rose JK, Adams GA, Callione CJ (1984) The presence of cysteine in the cytoplasmic domain of the vesicular stomatitis virus glycoprotein is required for palmitate addition. Proc Natl Acad Sci USA 81:2050–2054

Rose MD, Misra LM, Vogel JP (1989) KAR2, a karyogamy gene, is the yeast homolog of the mammalian BiP/GRP78 gene. Cell 57:1211–1221

Rothblatt JA, Webb JR, Ammerer G, Meyer DI (1987) Secretion in yeast: structural features influencing the post-translational translocation of prepro-alpha-factor in vitro. EMBO J 6: 3455–3463

Rothman JE (1987a) Protein sorting by selective retention in the endoplasmic reticulum and Golgi stack. Cell 50: 521–522

Rothman JE (1987b) Transport of the vesicular stomatitis glycoprotein to trans Golgi membranes in a cell-free system. J Biol Chem 262: 12502–12510
Rothman JE (1989) Polypeptide chain binding proteins: Catalysts of protein folding and related processes in cells. Cells 59: 591–601
Ruohola H, Kabcenell AK, Ferra-Novich S (1988) Reconstitution of protein transport from the endoplasmic reticulum to the Golgi complex in yeast: The acceptor Golgi compartment is defective in the sec23 mutant. J Cell Biol 107: 1465–1476
Sabatini BD, Kreibich, Morimoto T, Adesnik M (1982) Mechanisms for the incorporation of proteins in membranes and organelles. J Cell Biol 92: 1–22
Salminen A, Novick P (1987) A ras-like protein is required for a post-Golgi event in yeast secretion. Cell 49: 527–538
Saraste J, Kuimanen E (1984) Pre- and post-Golgi vacuoles operate in the transport of Semliki Forest virus membrane glycoproteins to the cell surface. Cell 38: 535–549
Saraste J, Palade GE, Farquhar MG (1986) Temperature-sensitive steps in the transport of secretory proteins through the Golgi complex in pancreatic cells. Proc Natl Acad Sci USA 83: 6425–6429
Schmidt MFG, Schlessinger MJ (1979) Fatty acid binding to vesicular stomatitis virus glycoprotein: a new type of post-translational modification of a viral protein. Cell 17: 813–819
Schmidt MFG, Schlessinger MJ (1980) Relation of fatty acid attachment to the translation and maturation of vesicular stomatitis and Sindbis virus membrane glycoproteins. J Biol Chem 255: 3334–3339
Schmitt HD, Puzicha M, Gallwitz D (1988) Study of a temperature-sensitive mutant of the ras-related YPT1 gene product in yeast suggests a role in the regulation of intracellular calcium. Cell 53: 635–647
Schultz AM, Henderson LE, Oroszlan S (1988) Fatty acylation of proteins. Ann Rev Cell Biol 4: 611–647
Segev N, Botstein D (1987) The ras-like yeast YPT1 gene is itself essential for growth, sporulation, and starvation response. Mol Cell Biol 7: 2367–2377
Segev N, Mulholland J, Botstein D (1988) The yeast GTP-binding protein YPT1 protein and a mammalian couterpart are associated with the secretion machinery. Cell 52: 915–924
Shelness GS, Kanwar YS, Blobel G (1988) cDNA-derived primary structure of the glycoprotein component of canine microsomal signal peptidase complex. J Biol Chem 263: 17063–17070
Siegel V, Walter P (1985) Elongation arrest is not a prerequisite for secretory protein translocation across the microsomal membrane. J Cell Biol 100: 1913–1921
Siegel V, Walter P (1988a) The affinity of signal recognition particle for presecretory proteins is dependent on nascent chain length. EMBO J 7: 1769–1775
Siegel V, Walter P (1988b) Each of the activities of signal recognition particle (SRP) is contained within a distinct domain: Analysis of biochemical mutants of SRP. Cell 52: 39–49
Silhavy TJ, Benson SA, Emr SD (1983) Mechanisms of protein localization. Microbiol Rev 47: 313–344
Sly WS, Fisher HD (1982) The phosphomannosyl recognition system for intracellular and intercellular transport of lysosomal enzymes. J Cell Biochem 18: 67–85
Stoller TJ, Shields D (1989) The propeptide of preprosomatostatin mediates intracellular transport and secretion of alpha-globin from mammalian cells. J Cell Biol 108: 1647–1655
Strous GJAM, Lodish HF (1980) Intracellular transport of secretory and membrane proteins in hepatoma cells infected by vesicular stomatitis virus. Cell 22: 709–717
Szczesna-Skopura E, Kemper B (1989) NH_2-terminal substitutions of basic amino acids induce translocation across microsomal membrane and glycosylation of rabbit cytochrome P450IIc2. J Cell Biol 108: 1237–1243
Talmadge K, Stahl S, Gilbert W (1980) Eukaryotic signal sequence transports insulin antigen in *Escherichia coli.* Proc Natl Acad Sci USA 77: 3369–3373
Towler DA, Gordon JI, Adams SP, Glaser L (1988) The biology and enzymology of eukaryotic protein acylation. Annu Rev Biochem 57: 69–99
Toyn J, Hibbs AR, Sanz P, Crowe J, Meyer DI (1988) In vivo and in vitro analysis of ptl1, a yeast ts mutant with a membrane-associated defect in protein translocation. EMBO J 7: 4347–4353
Trun NJ, Stader J, Lupas A, Kumamoto , Silhavy TJ (1988) Two cellular components, PrlA and SecB, that recognize different sequence determinants are required for efficient protein export. J Bacteriol 170: 5928–5930
von Heijne G (1983) Patterns of amino acids near signal-sequence cleavage sites. Eur J Biochem 133: 17–21
von Heijne G (1984) Analysis of the distribution of charged residues in the N-terminal region of signal

sequences: implications for protein export in prokaryotic and eukaryotic cells. EMBO J 3: 2315–2318

von Heijne G (1986) Towards a comparative anatomy of N-terminal topogenic protein sequences. J Mol Biol 189: 239–242

Walter P, Gilmore R, Blobel G (1984) Protein translocation across the endoplasmic reticulum. Cell 38: 5–8

Watanabe M, Blobel G (1989) Cytosolic factor purified from *Escherichia coli* is necessary and sufficient for the export of a preprotein and is a homotetramer of SecB. Proc Natl Acad Sci USA 86: 2728–2732

Waters MG, Chirico WJ, Blobel G (1986) Protein translocation across the yeast microsomal membrane is stimulated by a soluble factor. J Cell Biol 103: 2629–2636

Watts C, Wickner W, Zimmerman R (1983) M13 procoat and a pre-immunoglobulin share processing specificity but use different membrane receptor mechanisms. Proc Natl Acad Sci USA 80: 2809–2813

Weiss JB, Ray PH, Bassford PJ (1988) Purified SecB protein of *Escherichia coli* retards folding and promotes membrane translocation of the maltose-binding protein in vitro. Proc Natl Acad Sci USA 85: 8978–8982

Wiedmann M, Kurzchalia TV, Hartmann E, Rapoport TA (1987) A signal sequence receptor in the endoplasmic reticulum membrane. Nature 328: 830–833

Wieland FT, Gleason ML, Serafini TA, Rothman JE (1987) The rate of bulk flow from the endoplasmic reticulum to the cell surface. Cell 50: 289–300

Wills JW, Hardwick JM, Shaw K, Hunter E (1983) Alterations in the transport and processing of Rous sarcoma virus envelope glycoproteins mutated in the signal and anchor regions. J Cell Biochem 23: 81–94

Wills JW, Srinivas RV, Hunter E (1984) Mutations of the Rous sarcoma virus *env* gene that affect the transport and subcellular location of the glycoprotein products. J Cell Biol 99: 201–2023

Wilson C, Connolly T, Gilmore R (1988) Integration of membrane proteins into the endoplasmic reticulum requires GTP. J Cell Biol 107: 69–77

Wilson DW, Wilcox CA, Flynn GC, Chen E, Kuang W-J, Henzel WJ, Block MR, Ullrich A, Rothman JE (1989) A fusion protein required for vesicle-mediated transport in both mammalian cells and yeast. Nature 339: 355–359

Zagouras P, Rose JK (1989) Carboxy-terminal SEKDEL sequences retard but do not retain two secretory proteins in the endoplasmic reticulum. J Cell Biol 109: 2633–2640

Protein Sorting in Polarized Epithelial Cells

R. W. Compans and R. V. Srinivas

Department of Microbiology, University of Alabama at Birmingham, UAB Station, Birmingham, AL 35294, USA

Current Topics in Microbiology and Immunology, Vol. 170

1 Introduction

Epithelia are comprised of sheets or masses of cells, closely linked by junctional complexes which separate their plasma membranes into two domains, the apical and basolateral membranes. The epithelial cell layers from diverse tissues display morphological and functional polarity, often exhibit a high transepithelial electrical resistance, and transfer water and select solutes vectorially across the epithelium. Transport systems responsible for this vectorial transport are localized on either the apical or the basolateral membrane of the epithelial cells. Moreover, the apical and basolateral plasma membranes have distinct protein and lipid compositions, indicating that epithelial cells possess mechanisms for targeting proteins and lipids to specific plasma membrane domains. This review is focused on the process of protein sorting in polarized epithelial cells.

2 Properties of Epithelial Cells

2.1 Morphological Properties

Simple epithelia are composed of a single layer of epithelial cells that line a body cavity, or a luminal space such as the respiratory or gastrointestinal passage. The plasma membrane exposed to the luminal surface is generally described as the apical membrane. The apical membranes may be highly organized, and characterized by the presence of cilia (as in ciliated tracheal epithelia) or highly ordered microvilli (as in brush border epithelia of the intestine). In other tissues (for example, skin and urinary bladder), the epithelium may consist of several layers of cells, and is referred to as stratified epithelium. The liver mesenchyme displays a more complex organization, consisting of cuboidal hepatocytes with sinusoidal (analogous to basal) and canalicular (apical) membranes. Studies on the morphological and physiological characteristics of epithelia have involved use of whole organs, as well as isolated epithelial membranes or cultured epithelial cells. In most cases, the cultured cells appear to retain many properties of differentiated epithelium (TAUB 1985). Also, several continuous cell lines of epithelial origin have been described which retain the morphological and functional polarity observed with epithelial tissues, and studies with such cell lines have provided valuable information regarding morphological and biochemical aspects of epithelial cell polarity.

A common ultrastructural feature of epithelia is the presence of a tripartite junctional complex between adjacent cells (FARQUHAR and PALADE 1963). The elements of this tripartite complex are the tight junctions (*zonulaeoccludentes*), the intermediate junctions (zonulae adherentes), and the desmosomes (maculae adherentes). In addition, epithelial cells also contain gap or communicating junctions (REVEL and KARNOVSKY 1967) through which ions and small molecules can be exchanged between cells without access to the extracellular space. The cells also contain structures at their basal side described as hemidesmosomes, which serve as membrane-anchoring sites for tonofilament bundles that interconnect the desmosomes and hemidesmosomes. The hemidesmosomes form over specific sites on the basal lamina at the sites of insertion of anchoring fibrils (GIPSON et al. 1983).

Zonulae Occludentes. The tight junctions are characterized by apparent fusion of the external lamellae of adjacent cell membranes, resulting in obliteration of the intercellular space over variable distances. They appear to form a continuous belt-like attachment zone that serves as a permeability barrier between the apical and basolateral domains of the plasma membrane and the corresponding extracellular spaces (FARQUHAR and PALADE 1963). Freeze-fracture techniques have shown that the region morphologically referred to as a "tight junction" consists of a filamentous network composed of strands of intramembranous particles which span the thickness of the membranes (CLAUDE and GOODENOUGH 1973; MARTINEZ-PALOMO and ERLIJ 1975). It has been debated whether the tight junctional strands are composed of lipids or proteins. Based on th freeze-fracture and membrane fusion images, Kachar and colleagues (KACHAR and REESE 1982; PINTO DA SILVA and KACHAR 1982) have suggested that tight junctional strands represent cylindrical, inverted lipid micelles. On the other hand, there is also evidence to indicate that tight junctions may be comprised of proteins (GRIEPP et al. 1983; FEY et al. 1984). A monoclonal antibody directed against uvomorulin-like polypeptides specifically stained the junctional complex region of some epithelial cells (IMHOFF et al. 1983) and was found to disrupt normal cell polarity in the Madin-Darby canine kidney (MDCK) cell line (BEHRENS et al. 1985). More recently, STEVENSON et al. (1986) have obtained a monoclonal antibody which recognizes a polypeptide of ~225 kDa that is localized by immunoelectron microscopy precisely to the points of membrane contact in hepatocyte tight junctions. Antibody to this protein has also been shown to stain tight junctions from a variety of epithelia, including the MDCK epithelial cell line.

Zonulae Adherentes. The intermediate junctions are characterized by an intercellular space (~20 nm) occupied by homogeneous and apparently amorphous material. The adjoining plasma membranes are strictly parallel over distances of 200–500 nm, and conspicuous bands of dense material are located in the adjacent cytoplasmic matrix. The zonula adherens (also known as belt desmosome) is arranged as a belt-shaped junction just proximal to the tight junction. The electron-dense region possibly contains calcium-dependent adhesion-mediating proteins (BOLLER et al. 1985) and a 135-kDa protein (VOLK and GEIGER 1984). These junctions also contain a membrane-bound cytoplasmic plaque composed of vinculin, and possibly other peripheral proteins, which serves to anchor actin filaments, as well as a contractile actin filament bundle which forms a "terminal web" and connects the actin network in the cell to the cytoplasmic plaque of the junctional complex (HULL and STAEHLIN 1979; TILNEY

1983). The primary function of the *zonula adherens* seems to be mediation of cell-to-cell attachment. Recent studies have indicated that transfection of fibroblastoid tumor cells with a cDNA clone encoding liver cell adhesion molecules (L-CAM) will confer an epithelioid morphology, but not transepithelial resistance to the cells (MEGE et al. 1988), suggesting that L-CAM expression may be an early and important stage in epithelial differentiation.

Desmosomes. Desmosomes are most abundant in tissues prone to mechanical stress, e.g., the epithelium of skin, uterus, cervix, and bladder (McNUTT and WEINSTEIN 1973), and seem to assemble during development when strong and stable cell adhesion is required (DUCIBELLA et al. 1975). Desmosomal junctions are presumably involved in adhesion of adjacent plasma membrane domains by extracellular cross-bridging structures, which span 30-nm spaces between parallel plasma membranes. Bundles of tonofilaments, the most complex and highly developed class of intermediate filaments, terminate near or pass through the electron-dense submembranous plaques. As a major component of all epithelial cells, they appear in the electron microscope as filaments of intermediate length, aggregated laterally into anastamosing bundles and anchored at the cell periphery by looping into desmosomes (LANE 1982). This morphology has led to the suggestion that tonofilaments may play a central role in maintaining the tensile strength and integrity of epithelial cell sheets (LANE 1982).

2.2 Biochemical Properties of the Plasma Membranes

Differences in the compositions of the apical and basolateral membranes have been observed both in epithelial tissues and in cultured epithelial cells. A further distinction in chemical composition of the basal and lateral membrane domains has been reported in certain cell types such as hepatocytes (ROMAN and HUBBARD 1983). These differences are observed in lipids as well as proteins.

2.2.1 Lipids

Early studies on the lipid composition of brush border (apical) and basolateral membranes isolated from small intestines (FORSTNER et al. 1968; DOUGLAS et al. 1972; KAWAI et al. 1974; BRASITUS and SCHATER 1980) indicated quantitative differences in the lipid compositions of these membranes. The apical membranes showed a two- to fourfold enrichment in glycolipids and cholesterol, while the basolateral membranes were enriched in phosphatidylcholine. Both membranes contained similar amounts of phosphatidyl-ethanolamine. In certain cell types, the apical membranes were also enriched for sphingomyelin (MEIER et al. 1984; MOLITORIS and SIMON 1985). Quantitative differences have also been observed in the lipid composition of viruses that mature exclusively from apical (influenza virus) or basolateral (vesicular stomatitis virus) membranes of epithelial cells such as Madin-Darby bovine kidney (MDBK) and MDCK, while no such differences were observed in viruses grown on nonpolarized baby hamster kidney cells (KLENK and CHOPPIN 1970a, b; ROTHMAN et al. 1976; VAN MEER and SIMONS 1982).

The available information suggests that the differences in lipid composition are confined to the outer, exoplasmic leaflet of the plasma membrane bilayer, while the inner, cytoplasmic leaflet is contiguous throughout the cell. The tight junctions, by virtue of their barrier functions, are thought to maintain lipid polarity (DRAGSTEN et al. 1981, 1982; SPIEGEL et al. 1985, VAN MEER and SIMONS 1988). Consistent with this idea, fluorescent lipid probes introduced into the inner leaflet of the apical membrane can freely diffuse to the basolateral side, whereas probes introduced into the outer leaflet of the apical membrane remain confined to the apical plasma membrane (DRAGSTEN et al. 1981, 1982). The precise mechanisms involved in the generation of lipid asymmetry in the plasma membrane of epithelial cells are not clearly understood. The available evidence suggests that apical lipids are sorted from the basolateral lipids in an intracellular compartment (VAN MEER et al. 1987; VAN MEER and SIMONS 1988).

2.2.2 Proteins

Asymmetry in protein composition of apical and basolateral membranes of epithelia cells has been demonstrated by a variety of techniques. The apical or basolateral localization of enzymes, transport proteins, and other plasma membrane proteins has been addressed in detail in several recent reviews (MURER and KINNE 1980; KENNY and MAROUX 1982; RODRIGUEZ-BOULAN 1983; SABATINI et al. 1983; ALMERS and STIRLING 1984; SIMONS and FULLER 1985; RODRIGUEZ-BOULAN and NELSON 1989), and will be discussed only briefly here.

Nearly all the cell surface proteins examined have been found to be localized either on the apical or basolateral membrane domains of a given epithelial cell type. In addition, a protein localized at a particular plasma membrane domain in a given epithelial cell type is in general found in the same domain in other cell types. Enzymes which have been localized to apical membranes of various epithelial cells include Mg^{2+} ATPase (WISHER and EVANS 1975; GEORGE and KENNY 1973), neutral endopeptidase (DANIELSON et al. 1980; KERR and KENNY 1974) 5′-nucleotidase (COLAS and MAROUX 1980; MIEER et al. 1984; GEORGE and KENNY 1973), maltase (SEMENZA 1976; KERJASCHKI et al. 1984), leucine aminopeptidase (SEMENZA 1976; ROMAN and HUBBARD 1984a, b; DESNUELLE 1979), and γ-glutamyl transferase (MARATJE et al. 1979; INOUE et al. 1983). In contrast, adenylate cyclase (REIK et al. 1970; LODJA 1974 and SCHWARTZ et al. 1974) has been localized exclusively to the basolateral membranes of various epithelial cells. Localization of some proteins such as alkaline phosphatase, NA^+/K^+ ATPase, and the furosemide-sensitive sodium/potassium chloride cotransport protein is ambiguous. Furosemide-sensitive sodium/potassium chloride cotransport was localized to the apical domain in renal tubules (KEONIG et al. 1983) and shark rectal gland (HANNAFIN et al. 1983), but to the basolateral domain in corneal (LUDENS et al. 1980) and MDCK (ALTON et al. 1982) cells. Since the protein responsible for this activity has not been identified, it has been suggested that the observed differences may be due to the fact that this activity may be mediated by distinct gene products in different cell types, or due to the association of a homologous functional unit with distinct subunits that are sorted to different domains in different cell types (SIMONS and FULLER 1985). Alkaline phosphatase was localized to apical domains in placenta (CARLSON et al. 1976), pig kidney cell cultures (RABITO et al. 1984) and renal tubules (GOMORI 1941; HEIDRICH et al. 1972) but was found to be localized on basolateral domains of hepatocytes (WISHER and EVANS 1975; MEIER et al. 1984). Interestingly, in enterocytes alkaline phosphatase has been localized to the apical membranes by cytochemical techniques (LODJA 1974; BORGERS 1973), while MIRCHEFF et al. (1979) have found this activity in highly purified basolateral membrane fractions from rat duodenal cells. A similar discrepancy has also been encountered with the localization of Na^+/K^+ ATPase in choroid plexus cells (MILHORAT et al. 1975) and hepatocytes (TAKEMURA et al. 1984; MEIER et al. 1984). It is possible that different proteins with similar activities are being detected in these studies. For example, the Na^+/K^+ ATPase is a complex composed of α and β subunits, both of which may be immune-precipitated with monoclonal antibodies against either α and β subunit under nondenaturing conditions (QUARONI and ISSELBACHER 1981; KASHGARIAN et al. 1985). In immunofluorescence assays, monoclonal antibodies against both α and β subunits exclusively label the basolateral membranes of small intestinal and proximal colon epithelial cells. However, in distal colon epithelial cells, antibodies against the α subunit react exclusively with the basolateral membranes, while antibodies directed against the β subunit react with both basolateral membrane and the apical brush border (MARXER et al. 1989). These results indicate that in distal colon, unassembled β subunit of the Na^+/K^+ ATPase or an antigenically cross-reactive protein is transported to the apical membranes.

A class of membrane glycoproteins has been recently described in which the extracellular domain is linked to the membrane by a glycosyl phosphatidylinositol (GPI) linkage (LOW 1987; CROSS 1987; FERGUSSON and WILLIAMS 1988). Several enzymes which are anchored to membranes by a GPI linkage, including alkaline phosphatase (GOMORI 1941; LODJA 1974), renal dipeptidase (BARTH et al. 1974), and 5′-nucleotidase (COLAS and MAROUX 1980; MEIER et al. 1984), have been localized at apical membranes of various epithelial cell types. Six different polypeptide species were identified as GPI-anchored proteins on apical surfaces of MDCK cells, whereas no proteins anchored by this linkage were detected on basolateral membranes (LISANTI et al. 1988). In addition, expression of human Thy-1, a GPI-anchored protein, in transgenic mice was observed preferentially on apical surfaces of kidney epithelial cells (KOLLIAS et al. 1987). Thus, the available information indicates preferential localization of the GPI-linked class of membrane proteins on apical cell surfaces.

2.2.3 Extracellular Matrix

Epithelial and endothelial cells usually rest on a basement membrane or a basal lamina, which represents a special type of extracellular matrix containing type IV collagen, laminin, and proteoglycans, mainly of the heparan sulfate proteoglycan variety (KLEINMAN et al. 1981; TIMPLE and MARTIN 1982). The binding of the epithelial cell to the basement membrane is mediated via the hemidesmosomes. Additionally, the basolateral membranes contain specific receptors for collagen and laminin (SUGRUE and HAY 1981; VON DER MARK et al. 1984). A fully developed, morphologically distinguishable basal lamina may not be required for the establishment and maintenance of epithelial cell polarity. Epithelial cells adapted to grow in culture, e.g., MDCK or murine mammary epithelial cells, can form epithelial sheets with no apparent basal lamina. Also, a basal lamina is usually not observed beneath the basolateral membranes of hepatocytes in most mammalian livers. Nevertheless, secretion of small amounts of extracellular matrix (ECM) components by these cells may be necessary and sufficient for the establishment of polarity. Indeed the growth of certain epithelial cells in culture seems to require ECM components (CHAMBARD et al. 1981).

2.3 Permeability Characteristics

The apical and basolateral membranes of epithelial cells have distinct ion pumps and generate a high ionic gradient across the epithelial monolayer. The tight junctions form an efficient barrier for the diffusion of ions and macromolecules, resulting in a high transepithelial electrical resistance. There is considerable heterogeneity in the degree of resistance displayed by different epithelia both from tissue and in cell culture. For example, rabbit ileum has a transepithelial resistance of $125\,\Omega/cm^2$, whereas urinary bladder has a resistance of $12\,000\,\Omega/cm^2$ (REUSS and FINN 1974). Two strains of MDCK cells, strain I with high ($>1000\,\Omega/cm^2$) and strain II with low ($\sim 300\,\Omega/cm^2$) resistance, have been described (RICHARDSON et al. 1981; FULLER and SIMONS 1986). Epithelia with low resistance display a greater transepithelial permeability to substances such as mannitol, inulin, and sucrose and have therefore been described as "leaky" epithelia. In accordance with the view that these molecules do not enter cells readily, if at all, it is likely that leakiness of epithelia in general is the result of paracellular passage of molecules.

Early studies with a variety of unperturbed native epithelia indicated that specific structural aspects of the tight junctions, such as the number of strands, or depth and complexity of their network assessed from freeze-fracture replicas, correlated with the ability of epithelia to resist passive transepithelial ion flow (CLAUDE and GOODENOUGH 1973; CLAUDE 1978). It was also observed that tight junction strand counts correlate positively with paracellular resistance. However, the comparison of tight junction structure and paracellular resistance of epithelial linings from mammalian ileum and toad bladder failed to support this view (MARTINEZ-PALOMO and ERLIJ 1975). MADARA and DHARMSATHAPHORN (1985) have observed that when tight junction heterogeneity exists in a monolayer, the mean values of structure or function may not correlate well with the net behavior of the population. Studies with MDCK cells have indicated that current sinks exist in a monolayer with low resistance and may be localized to a minor subpopulation of intercellular junctions with a low strand count (CEREIJIDO et al. 1980).

Various perturbants of epithelial permeability, such as removal of Ca^{2+} ions by EGTA, result in the opening of the tight junctions. The effect of these perturbations is generally reversible, and accompanied by resealing after removal of the perturbing stimuli. This reversible modification in the permeability of the tight junctions occurs even in the presence of inhibitors of protein synthesis, indicating that resealing is a consequence of junctional reassembly rather than de novo synthesis of the membrane junctional components (MARTINEZ-PALOMO et al. 1980). Available evidence indicates that cytoskeletal elements may play some role in the resealing of opened tight junctions. Cytoplasmic microtubules form a dense network of fibers that extend radially to the periphery of epithelial cells. Colchicine treatment leads to microtubule depolymerization and results in the disruption of the tubular network, leaving remnants of microtubules in contact with the plasma membrane at the periphery, and the centrosome in the perinuclear region (MEZA et al. 1980).

2.4 Plasticity of Epithelial Polarity

Several lines of evidence indicate that, under appropriate conditions, the polarity of epithelial, organization may be reversed or reoriented. JAFFE (1981) and STERN and MACKENZIE (1983) demonstrated that reversal of electrical potential across a transporting epithelium results in a shift in the position of the tight junctions; also, the transporting epithelia were found to pump Na^{+} after imposition of the electrical potential. Morphological and electrochemical evidence has also been obtained to show that ion pumps in renal epithelia can reorient in response to induced electrolyte imbalance in intact animals. In other examples, isolated inverted follicles of thyroid cells, with apical membranes facing outward, have been shown to undergo a reversal of polarity to form follicles with apical membranes facing the lumen when they are embedded in a collagen matrix (CHAMBARD et al. 1981). This process is thought to involve recruitment of the collagen receptor to the apical membrane and its interaction with ECM to form a basal pole and subsequent reorganization of the epithelial architecture. The molecular and cellular mechanisms involved in this reorientation of epithelial polarity are unclear. It remains to be established whether all the biochemical and immunochemical parameters of the epithelial cells are reoriented during such reversal of epithelial polarity.

3 Transport of Secreted Proteins

3.1 Secretion of Endogenous and Foreign Proteins in MDCK Cells

In general, proteins secreted by epithelial cells are targeted to either their apical or basolateral surfaces. In some cell types, especially those from exocrine and endocrine glandular tissues, secretory proteins are packaged into specialized vesicles, termed secretory granules, that fuse with specific membrane domains after an appropriate stimulus, resulting in regulated secretion (BURGESS and KELLY 1987). In many instances, e.g., thyroid follicular cells, which are organized into cavitary structures, or pancreatic endocrine cells (LOMBARDI et al. 1985), these cell types also display plasma membrane polarity, and the contents of the secretory granules are released in a polar manner. These cells also possess a constitutive pathway for secretion, which does not involve packaging in secretory granules. Available evidence indicates that in certain endocrine tissues, different hormones are packaged into distinct secretory granules (HASHIMOTO et al. 1987), suggesting additional complexity in the sorting machinery. However, other cell types, including MDCK cells, do not possess secretory granules; nevertheless, the proteins which are secreted constitutively from these cells are targeted to specific plasma membrane domains. MDCK cells secrete a major 80-kDa protein complex into their apical medium, whereas the secretion of the basement membrane components laminin and heparan sulfate proteoglycan (HSPG) take place at basolateral surfaces (KONDOR-KOCH et al. 1985; GOTTLIEB et al. 1986a; CAPLAN et al. 1987).

One possible mechanism for transport of secretory proteins to the cell surface is transport by default, i.e., proteins lacking sorting signals would be incorporated randomly into transport vesicles and carried to the cell surface by bulk flow. Evidence in favor of transport by a default pathway was obtained by adding tripeptides with the sequence Asn-X-Ser/Thr, representing potential glycosylation sites, to cells; these peptides were reported to be internalized, glycosylated, and rapidly secreted (WIELAND et al. 1987). Since it is unlikely that a tripeptide could possess a specific sorting signal, these results indicated that transport of these peptides occurs by a bulk flow process. Two lines of evidence support the conclusion that MDCK cells possess a default pathway for secretion which is non-directional, i.e., that proteins lacking signals for directional transport are released both apically and basolaterally. Several genes which encode endocrine proteins were introduced into MDCK cells by transfection (KONDOR-KOCH et al. 1985; GOTTLIEB et al. 1986b). The resulting proteins, which are not normally expressed in MDCK cells, were observed to be released from both the apical and basolateral surfaces, under conditions where the endogenously secreted proteins were targeted to a specific membrane domain. It was postulated that these exogenous proteins were not recognized by the sorting mechanisms for endogenous proteins in MDCK cells, and were thus released nondirectionally by a default process. A second approach to identify the default pathway was based on the observation that in cells treated with NH_4Cl, a weak base that raises the pH of intracellular vesicles, lysosomal enzymes are unable to interact with their receptors and are released into the culture medium (Brown et al. 1984). In MDCK cells treated with NH_4Cl, the lysosomal enzyme cathepsin D was released into both apical and basolateral medium (Caplan et al. 1987). No release of the enzyme was observed in the absence of NH_4Cl. Treatment of cells with NH_4Cl also affected the sorting of laminin and HSPG in MDCK cells, resulting in their nondirectional secretion (Caplan et al. 1987). In contrast, the apical secretion of the 80-kDa protein complex was unaffected. The investigators suggested that the transport pathway of proteins destined for basolateral secretion involves an acidic compartment, and that treatment with weak bases disrupts recognition by the sorting mechanism in such acidic vesicles.

Taken together, these observations indicate that proteins which are secreted constitutively at either the apical or basolateral membrane domains of MDCK cells are recognized by specific sorting mechanisms, whereas proteins lacking such recognition signals may be incorporated by default into vesicles destined for transport either to apical or basolateral membranes. The molecular features of secretory proteins which may serve as recognition signals for sorting are not known. A likely mechanism for the sorting of secretory proteins involves their interaction with a membrane-bound receptor, which is itself targeted to a specific destination (secretory granules, apical or basolateral plasma membranes). Evidence in support of such a mechanism was obtained in a study of the osteoclast, a polarized cell which secretes large amounts of lysosomal enzymes into an apical lacuna (Baron et al. 1988). It was found that in this cell type the mannose-6-phosphate receptors for lysosomal enzymes are codistributed with lysosomal enzymes along the exocytic pathway to the apical membranes, suggesting that these receptors are involved in targeting of the enzymes for apical secretion. Such receptors presumably are also present for proteins secreted endogenously in a directional manner, such as the 80-kDa proteins secreted from apical cell surfaces of MDCK cells. In contrast, cells may not possess receptors for foreign secretory proteins, with the consequence that such proteins would be incorporated into transport vesicles at random and released by a default process.

3.2 Tissue-Specific Differences in Sorting of Secretory Proteins

In contrast to the nondirectional secretion of foreign proteins by MDCK cells, it was reported that in a line of human intestinal epithelial cells, CaCo-2, such foreign glycoproteins were secreted exclusively at basolateral surfaces (Rindler and Traber 1988). Further, all the radiolabeled proteins endogenously secreted by these cells were also released exclusively at basolateral surfaces, and treatment of these cells with weak bases resulted in the release of lysosomal enzymes primarily at basolateral surfaces. Thus it appears that these intestinal cells possess a mechanism for directing all secretory proteins exclusively to their basolateral membranes. Similarly, hepatocytes apparently possess only a basolateral pathway for protein secretion (BARTLES and HUBBARD 1988). Nevertheless, CaCo-2 cells appeared to possess a normal pathway for apical transport of membrane glycoproteins (RINDLER and TRABER 1988). These results lead to the conclusions that there are important cell-type-dependent differences in the sorting of secretory proteins and that differences also exist between the sorting mechanisms for secretory vs membrane proteins in the same cell type.

Evidence has been obtained that secretory proteins and membrane proteins can be localized in the same intracellular membrane vesicles (STROUS et al. 1983; BURGESS and KELLY 1987). Although this has not yet been reported for polarized epithelial cells, it is likely that vesicles delivering membrane proteins to apical or basolateral surfaces would also contain proteins destined for nontargeted secretion in their contents, unless specific mechanisms exist for their exclusion from such vesicles. No information has been reported on the relative amounts of vesicular traffic involved in transport of apical vs basolateral membranes in the CaCo-2 cell line, but in vivo studies indicate that a significant fraction of newly synthesized membrane proteins in intestinal epithelial cells are targeted to apical membranes (BLOK et al. 1981; BENNET et al. 1984). If such proteins are directly delivered by transport vesicles destined for apical membranes, as occurs in MDCK cells (see below), the secretion of proteins exclusively from basolateral surfaces of intestinal cells requires a mechanism for exclusion of secretory proteins from the vesicles involved in transport of apical membrane proteins.

3.3 Secreted Forms of Membrane Proteins

Truncated forms of several membrane glycoproteins have been produced by expression of genes in which coding regions for membrane-anchoring sequences have been deleted by recombinant DNA techniques, resulting in proteins that are efficiently secreted into culture media. A soluble, truncated form of the murine retrovirus envelope protein was found to be secreted nondirectionally into apical and basolateral media of MDCK cells, in contrast to the membrane-anchored protein which is expressed only on basolateral membranes (STEPHENS and COMPANS 1986). Similarly, truncated forms of the influenza hemagglutinin (HA) glycoprotein and the vesicular stomatitis virus (VSV) spike glycoprotein (G protein) were reported to be released nondirectionally from transfected MDCK cells (GONZALEZ et al. 1987). In contrast to these observations, it was reported that a secreted form of the influenza HA glycoprotein was preferentially released at apical surfaces of the MA104 line of polarized monkey kidney cells (ROTH et al. 1987). The reason for the difference between this result and those mentioned above remains to be determined. It is possible that the sorting mechanism in MA 104 cells, but not in MDCK cells, is able to recognize a signal in the luminal domain of HA, or that the truncated HA molecules are released by default pathways in both cell types but that the default pathway in MA 104 cells preferentially releases secreted proteins at the apical surface. With the exception of this report, the results indicate that the mechanisms which normally target membrane proteins to specific domains are no longer able to recognize the secreted forms of these proteins, and suggest that they are released by a nondirectional default pathway which may be the same as that used by foreign secretory proteins. It is possible that the sorting signals have been deleted from the truncated proteins, or alternatively that membrane anchorage per se is a requirement for recognition of the sorting signals.

4 Transport of Membrane Proteins

4.1 Polarized Maturation of Viruses

Enveloped viruses, which acquire a lipid-containing membrane during the process of assembly by budding at a cellular membrane, have been used extensively to study the biosynthesis and transport of membrane proteins. The structure and assembly of lipid-containing viruses have been reviewed in detail elsewhere (e.g., COMPANS and KLENK 1979; SIMONS and GAROFF 1980; DUBOIS-DALCQ et al. 1984; STEPHENS and COMPANS 1988). The virions contain one or more virus-coded surface glycoproteins embedded as transmembrane proteins in a host-cell-derived lipid bilayer. Many enveloped viruses are assembled by budding at the plasma membrane, although others are assembled by budding at internal membranes (see Sect. 5).

Studies of the transport of membrane proteins in polarized epithelial cells were greatly stimulated by the finding that enveloped viruses mature by budding at specific plasma membrane domains in such cells (RODRIGUEZ-BOULAN and SABATINI 1978). In polarized MDCK cells, influenza virus and paramyxoviruses were found to mature exlusively at apical membranes, whereas VSV and C-type retroviruses are assembled only at basolateral membranes (RODRIGUEZ-BOULAN and SABATINI 1978; ROTH et al. 1983a). Examples of the polarized budding of influenza virus and VSV are shown in Fig. 1. The surface glycoproteins of each of these viruses are localized on the same membrane domain where virus maturation occurs–(RODRIGUEZ-BOULAN and PENDERGAST 1980; ROTH et al. 1983b). Similar observations were made when these viruses were used to infect other polarized epithelial cell types, indicating that a variety of epithelial cells possess similar mechanisms for targeting proteins to apical or basolateral membranes, and that viral glycoproteins are recognized by these sorting mechanisms. The glycoproteins of many enveloped viruses are synthesized at high levels in infected cells, the molecules are well characterized, and the three-dimensional structures are known for two of them, as discussed below. Thus, virus-infected epithelial cells have become the system of choice in a number of laboratories for investigation of the mechanisms involved in sorting of membrane glycoproteins. Table 1 summarizes the information which has been obtained about sites of virus maturation and viral glycoprotein expression in polarized epithelial cells.

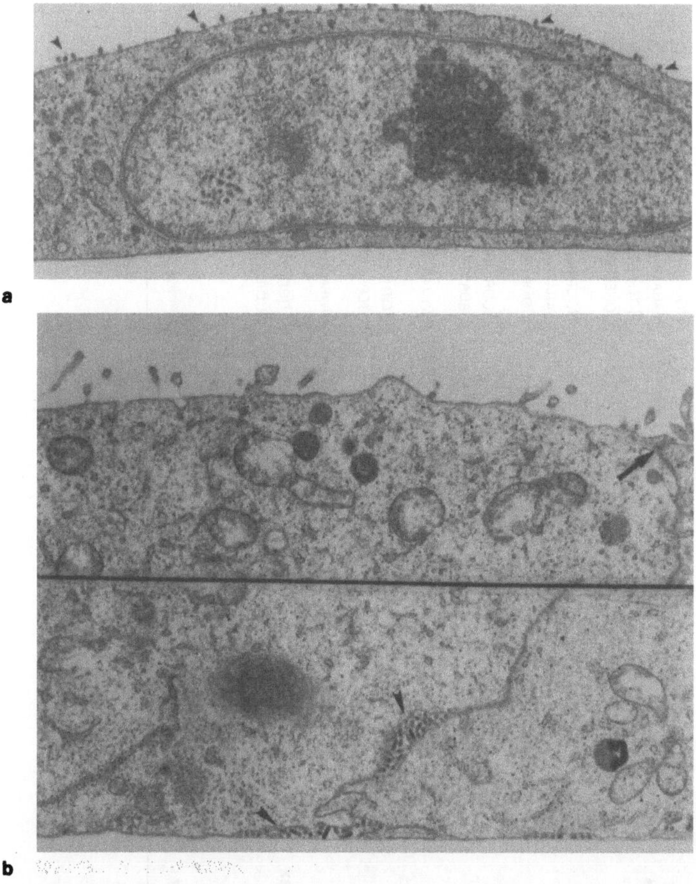

Fig.1 a, b. Polarized budding of enveloped viruses in MDCK cells. **a** Low-magnification view of an influenza-virus-infected MDCK cell, showing virions in the process of budding at the apical membrane (*arrowheads*). The basolateral membrane (below the cell) is devoid of virions (× 11 000). **b.** Portions of the apical (*top panel*) and basolateral (*bottom panel*) membranes of a VSV-infected MDCK cell, showing virions released at the basal and lateral membranes (*arrowheads*). No virus is seen at the apical surface. The *arrow* points to a fight junction. (× 18 000)

The influenza A viruses encode two major glycoproteins, the HA and the neuraminidase (NA), both of which are expressed on apical surfaces of polarized cells (ROTH et al. 1983b; JONES et al. 1985). The influenza HA is synthesized as a precursor, approximately 77 kDa in size, which is proteolytically cleaved into two polypeptide chains, designated HA1 (~50 kDa) and HA2 (~27 kDa), which are held together by disulfide bonds (LAZAROWITZ et al. 1971; SKEHEL 1972). The HA spike is a noncovalently linked trimer (WILEY et al. 1977; WILSON et al. 1981). Near the C terminus (GETHING and SAMBROOK 1982) of HA2, a stretch of 24 hydrophobic residues is thought to anchor HA in the membrane. Treatment with the protease bromelain removes HA from the surfaces of influenza virions (COMPANS et al. 1970). The bromelain-cleaved HA (BHA) is soluble and has been crystallized, and its structure has been analyzed at a resolution of 3 nm (WILSON

Table 1. Sites of virus maturation and viral glycoprotein insertion in polarized epithelial cells

Family	Virus	Site of virus maturation	Glycoprotein	Glycoprotein localization	References
Enveloped RNA viruses					
Orthomyxoviridae	Influenza	Apical	HA	Apical	Rodriguez-Boulan and Sabatini (1978) Roth et al. (1983a)
			NA	Apical	Jones et al. (1985)
Para-myxoviridae	SV5, Sendai, measies	Apical	HN and F	Apical	Rodriguez-Boulan and Sabatini (1978) Compans et al. (1982)
Rhaboviridai	VSV	Basolateral	G	Basolateral	Rodriguez-Boulan and Sabatini (1978)
Togaviridae	SFV	Bosolateral	E1	Basolateral	Roman and Garoff (1986)
Retroviridae	MuLV	Basolateral	gp70/p15E	Basolateral	Roth et al. (1983b) Stephens et al. (1986)
Lentivirus	HIV	Basolateral	gp120/gp41	Basolateral	Owens et al. (1989)
Bunyaviridae	Punta Toro	Golgi complex	G1, G2	Golgi complex	Chen et al. (submitted for publication)
Enveloped DNA viruses					
Herpetoviridae	HSV-1 and 2	Inner nuclear	gB, gC, gD,	Basolateral	Srinivas et al. (1986)

et al. 1981; WILEY and SKEHEL 1987). It is an elongated cylinder consisting of a 76-nm-long fibrous stem (derived primarily from HA2) and a distal globular head (derived from HA1). The fibrous stem contains two antiparallel α helices, which extend from and terminate near the membrane in a globular fold consisting of five stranded, compact, antiparallel β sheets. The distal globular region of HA1 is connected to the HA2 fibrous stem by only two antiparallel chains.

The NA glycoprotein is a mushroom-shaped spike, consisting of a stalk and a head, and consists of tetramers of NA polypeptides (LAVER and VALENTINE 1969). The deduced amino acid sequence of NA reveals a single hydrophobic region near the N terminus which serves as a membrane anchor (FIELDS et al. 1981; BLOK et al. 1982). A small stretch of six N-terminal amino acids (residues 1–6) is thought to lie on the cytoplasmic side of the membrane, while hydrophobic residues from residues 7–35 span the membrane as an uncleaved signal anchor. Residues 36–73 are heavily glycosylated and constitute the stalk structure to which the heads are attached. The NA heads can be released as intact, soluble, and biologically active structures upon pronase digestion of intact virus particles. Pronase-solubilized NA heads have been crystallized and their structure determined at a resolution of 2.9 nm VARGHESE et al. 1983). The crystallographic data indicate a tetramer with a box-shaped (100 × 100 × 60 nm) head attached to a slender stalk. Each monomer consists of six topologically indentical, four-stranded, antiparallel β sheets which are arranged like the blades of a propeller.

The G protein of VSV is a prototype basolateral membrane protein, and its transport and biosynthesis have been investigated in great detail. The G protein monomer has a molecular mass of ~65 kDa and contains two N-linked oligosaccharide side chains. The primary structure of the G protein (ROSE and GALLIONE 1981; WAGNER 1987) reveals a polypeptide of 511 amino acids with an N-terminal cleavable signal sequence, a C-terminal hydrophobic membrane-spanning domain of 20 amino acids, and a hydrophilic cytoplasmic domain of 29 amino acids. Available evidence indicates that the VSV G protein exists as an oligomer (possibly trimer), although the detailed three-dimensional structure of the molecule has not been determined.

The primary structures of several other viral glycoproteins expressed on basolateral surfaces have also been determined. The members of the *Alphavirus* genus encode three envelope polypeptides designated E1, E2, and E3. E2 and E3 are cleavage products of a precursor protein p62 and contain 422 and 66 amino acids respectively (GAROFF et al. 1982). The membrane proteins p62 and E1 both contain N-terminal signal sequences and C-terminal hydrophobic membrane-anchoring sequences, such that 31 amino acids of p62 and two amino acids of E1 are exposed on the cytoplasmic side of the membrane. The two proteins exist as a complex in the endoplasmic reticulum (ER), and the cleavage of p62 into E2 and E3 occurs later in a post-Golgi step (GAROFF et al. 1977, 1980). The murine leukemia viruses (MuLV), members of the C-type retroviruses, code for a major envelope glycoprotein, which is first synthesized as a precursor polyprotein and subsequently cleaved into two subunits designated gp70 and p15E; the latter is further cleaved into p12E and a peptide designated as R during virus assembly. The precursor contains a cleavable N-terminal signal sequence, a C-terminal membrane anchor consisting of 29 hydrophobic residues and a 32-residue-long hydrophilic cytoplasmic tail. The overall organization of the envelope proteins in other retroviruses is similar to that in MuLV, but there are variations in the sizes of the external and transmembrane proteins (reviewed by WEISS et al. 1984).

The extensive information that has been obtained on the structure of viral glycoproteins has not yet revealed features which might serve as signals for targeting of the proteins to apical vs basolateral plasma membrane domains. Neither the transmembrane topology nor the size or charge of specific protein domains can be correlated with the site of surface expression, nor have any conserved sequences been identified that are shared among all apical or basolateral viral proteins.

4.2 Intracellular Events in Protein Sorting

4.2.1 Intracellular Sorting

Glycoproteins destined for apical or basolateral membranes both follow the exocytic pathway: they are synthesized on membrane-bound polyribosomes in the rough endoplasmic reticulum (RER) and transported through the Golgi complex *en route* to the cell surface. When MDCK cells are simultaneously infected by two different viruses with opposite budding polarity, such as influenza virus and VSV, budding of each virus continues to be restricted to specific membrane domains until late times post infection when junctional complexes are disrupted (ROTH and COMPANS 1981; RINDLER et al. 1984). Therefore, cells doubly infected by such viruses provide a useful system to establish the stage in the transport pathway at which sorting of glycoproteins occurs. In cells doubly infected with influenza virus and VSV, the glycoproteins of both viruses were localized in the same Golgi cisternae by immunoelectron microscopy (RINDLER et al. 1984), suggesting that the sorting process occurs during or after exit from the Golgi complex. Similar conclusions were obtained using a biochemical approach (FULLER et al. 1985).

A striking differential effect of the sodium ionophore, monensin, was observed on transport of glycoproteins of influenza virus and VSV in MDCK cells (ALONSO and COMPANS 1981). Transport of the VSV G proteins to basolateral cell surfaces was completely inhibited, whereas transport of influenza virus glycoproteins to apical surfaces and the assembly of influenza virions was unimpaired, despite marked dilation of the Golgi complex by monensin. Since monensin interferes with protein transport by impeding exit from the Golgi complex (TARTAKOFF and VASSALI 1977), these results suggested that significant differences exist in the process of exit of these apical and basolateral proteins from the Golgi complex. The ionophore treatment was found to partially block the oligosaccharide processing of the influenza HA protein, but such proteins were transported to the cell surface and incorporated efficiently into fully infectious virus particles (ALONSO-CAPLEN and COMPANS 1983). Since oligosaccharide processing was found to occur for the VSV glycoprotein in monensin-treated cells, these results suggested that influenza HA may bypass certain glycosyl transferases in the presence of monensin.

Following transport through the Golgi complex, viral glycoproteins are apparently sorted into distinct sets of transport vesicles which deliver them to a specific plasma membrane. Several laboratories, using a number of different experimental approaches, have obtained convincing evidence that viral glycoproteins are transported directly from the Golgi complex to their final plasma membrane destination in MDCK cells (MATLIN and SIMONS 1984; MISEK et al. 1984; PFEFFER et al. 1985; RINDLER et al. 1985). Thus, the sorting of membrane proteins in MDCK cells occurs intracellularly. Using a temperature-sensitive mutant to synchronize transport of the influenza HA protein between the Golgi complex and the cell surface, it was observed that the protein appeared to be transported in smooth membrane vesicles about 200–300 nm in diameter (RODRIGUEZ-BOULAN et al. 1984). These vesicles were not reactive with antibody to clathrin, indicating that clathrin-coated vesicles are not involved in transport. Evidence for intracellular sorting and polarized delivery to basolateral membranes has also been obtained for an endogenous membrane protein of MDCK cells, the α subunit of the Na^+, K^+ ATPase (CAPLAN et al. 1986). Thus, the sorting events for viral and cellular membrane proteins in MDCK cells appear to be similar processes.

Evidence has been obtained that the Golgi complex is an acidic compartment, resembling the components of the endocytic pathway in this regard (MATLIN 1986a, b). Treatment of cells with weak bases such as ammonium chloride or chloroquine, which raise the pH of compartments such as endosomes and lysosomes, has been used to investigate the role of intracellular acidic compartments. The effects of weak bases on the sorting of membrane and secretory proteins has been examined in MDCK cells as well as other polarized cell types. No effect was observed on the delivery of the influenza HA proteins to apical cell surfaces (MATLIN 1986a) or on the expression of Na^+/K^+ ATPase on basolateral membranes of MDCK cells (CAPLAN et al. 1986). In contrast, significant effects on polarized secretion of proteins were observed when weak bases were applied to MDCK cells (see Sect. 3.1). These differences may reflect fundamental differences in the sorting mechanisms of membrane-bound vs secreted proteins. The results suggest that surface charge may affect the interaction of secretory proteins with membrane-bound receptors involved in sorting, but that such charge effects do not play a role in the recognition of sorting sequences in the membrane proteins which have been studied.

4.2.2 Effects of Alteration of the Cytoskeleton or Cell Anchorage

Two groups have investigated the effects of agents which disrupt cytoskeletal elements upon polarized transport of glycoproteins and maturation of viruses in

MDCK cells (SALAS et al. 1986; RINDLER et al. 1987). No effect was observed when cells were treated with cytochalasin D, which disrupts action microfilaments. Microtubule-disrupting drugs—colchicine, taxol, or nocodazole—were found to have no effect on the basolateral expression of the VSV G protein or the maturation of VSV at basolateral membranes. SALAS et al. (1986) also reported that these agents had no effect on the transport of influenza HA to apical membranes or on the apical maturation of influenza virions under conditions where microtubules were disrupted. In contrast, RINDLER et al. (1987) reported that similar concentrations of these microtubule-disrupting agents had a significant effect on the influenza HA protein, resulting in its nonpolarized expression as well as in the budding of a major fraction of influenza virions at basolateral surfaces. Thus, the available findings concerning the possible role of microtubules as a determinant of surface polarity are inconsistent, and further studies are needed to resolve these apparent differences.

Transport of cellular proteins to apical membranes of intestinal epithelial cells is also apparently altered by agents that disrupt microtubules (BENNET et al. 1984; ELLINGER et al. 1983; PAVELKA et al. 1983). Nocodazole was reported to inhibit the transport of an apical membrane protein, aminopeptidase N, in CaCo-2 cells and to lead to its partial misdirection to the basolateral surface (EILERS et al. 1989). In contrast, no effect was observed on the expression of a basolateral membrane protein. The apical but not the basolateral secretion of two lysosomal enzymes was also blocked by nocodazole, indicating that an intact microtubule network is required for apical transport of secretory as well as membrane proteins, but not for basolateral transport of such proteins. It was suggested by EILERS and coworkers (1989) that disruption of apical transport could occur by one of two alternative mechanisms: treatment with microtubule-disrupting drugs has been reported to lead to fragmentation of the Golgi apparatus, which could disrupt transport; alternatively, transport vesicles might require intact microtubules for their directional transport to apical cell surfaces.

When virus-infected MDCK monolayers are dissociated into single cells by trypsin treatment, polarity of virus maturation is no longer observed (ROTH and COMPANS 1981; RODRIGUEZ-BOULAN et al. 1983). Similarly, cellular membrane proteins, which are localized to specific domains in intact epithelial monolayers, become uniformly distributed on surfaces of cells in which junctional complexes are disrupted (PISAN and RIPOCHE 1976; ZIOMECK et al. 1980). These observations indicate that integrity of the junctional complexes is required for maintenance of the polarized distribution of membrane proteins, and polarized budding of viruses. However, it has been observed that virus maturation is polarized in single cells sparsely plated on a collagen substrate, as well as in clusters of cells in suspension (RODRIGUEZ-BOULAN et al. 1983). In the latter case VSV maturation was observed in the intercellular spaces, corresponding to the basolateral domain, whereas influenza viruses was released from free cell surfaces. Therefore, the complete junctional complexes present in an intact cell monolayer are apparently not required for polarized virus maturation, and it appears that contact of a cell with the substrate or another cell is necessary and sufficient for the establishment of the distinct plasma membrane domains.

4.2.3 Basolateral-to-Apical Transport

In contrast to the intracellular sorting and directional transport of membrane proteins in MDCK cells, an alternative sorting pathway has been proposed in other cell types. Proteins destined for the apical membrane are thought to be first transported to the basolateral surface, and are subsequently translocated to the apical surface. Evidence for such a sorting process has been reported for intestinal epithelial cells (HAURI et al. 1979; QUARONI et al. 1979a, b; QUARONI and ISSELBACHER 1981) and hepatocytes (EVANS 1980; BARTLES et al. 1987). These studies were carried out using cell fractionation procedures combined with pulse-chase labeling, and it is difficult to rule out the possibility that some vesicles involved in intracellular transport of apical membrane proteins might be present in basolateral-membrane-containing fractions. Using a novel assay which combines metabolic labeling with domain-selective biotinylation (LISANTI et al. 1989) to monitor the transport of newly synthesized membrane proteins in epithelial cells, LEBIVIC et al. (1989) observed vectorial targeting of apical and basolateral membrane proteins in SK-CO-15 cells, a continuous line of colon epithelial cells derived from a human adenocarcinoma. These results need to be extended to other cell types and other proteins such as aminopeptidase, sucrose isomaltase, and dipeptidyl peptidase, which have been postulated to arrive at the basolateral membranes prior to their transport to apical membranes of intestinal epithelial cells. MATTER et al. (1990) recently observed that sucrose isomaltase was delivered directly to apical membranes from trans-Golgi network (TGN) of CaCo-2 cells, while a fraction of aminopeptidase and dipeptidyl peptidase arrived at the basolateral membranes prior to endocytosis and resorting to the apical membranes. It is therefore possible that sorting of apical proteins from the TGN may be less efficient in these cells, and the missorted molecules are efficiently resorted to the apical membranes by transcytosis.

A transcellular transport process has been demonstrated for membrane proteins involved as receptors for translocation of certain ligands across an epithelial layer. The best characterized of these is the receptor for transport of secretory immunoglobulin A (sIgA) across the epithelial cell layers of the intestine or the hepatocyte. The receptor for sIgA is a transmembrane protein that is initially transported to basolateral membranes after its synthesis (MOSTOV and BLOBEL 1983). Production of sIgA occurs in plasma cells which underly the intestinal epithelia. After binding of sIgA, the receptor-ligand complex undergoes endocytosis and is transported to the apical cell surface. Release of sIgA at the apical surface occurs following cleavage of the receptor; after cleavage, a fragment of the receptor designated secretory component (SC) remains associated with each antibody molecule (reviewed by MESTECKY and MCGHEE 1987). Thus, the transcellular transport process consumes one receptor molecule for each molecule of ligand which is transported. A similar process of receptor-mediated transcytosis occurs in hepatocytes (HOPPE et al. 1985). Thus, a basolateral-to-apical transport pathway is observed at least for those membrane proteins involved in binding and transport of macromolecules across epithelial cell layers. Since it is likely that tight junctions will serve to restrict lateral diffusion and therefore prevent extensive transcellular movement of proteins by lateral diffusion in the plane of the membrane, such redistribution of proteins from basolateral to apical surfaces presumably occurs by vesicular transport.

When a cDNA clone for the sIgA receptor was expressed in MDCK cells, which normally do not produce this receptor, the receptor molecules were transported selectively to the basolateral surface (MOSTOV and DEITCHER 1986). The receptor was found to be cleaved, resulting in the release of SC exclusively at the apical surface. This cleavage and release was observed whether or not the ligand, sIgA, was present in the culture medium. Thus it appears that the receptor possesses signals for targeting to basolateral membranes, but that after endocytosis and cleavage it is redirected to the apical surface. The redirection of the protein to the apical surface presumably involves the exposure of a dominant signal for transport to apical membranes, possibly as a result of acidification and/or cleavage, after endocytosis at basolateral membranes. Recent studies indicate that phosphorylation of a serine residue located on the cytoplasmic tail of the sIgA receptor is required for transcytosis, but not basolateral targeting (CASANOVA et al. 1990). Substitution of serine 664 with an aspartic acid residue also facilitated transcytosis, suggesting that a negative charge at that position is required for transcytosis.

4.3 Localization of Sorting Signals in Membrane Glycoproteins

Studies of the expression of cDNA clones of viral glycoprotein genes have shown that glycoproteins of several enveloped viruses are targeted specifically to the same membrane domain where virus assembly occurs, in the absence of any other viral proteins (ROTH et al. 1983a; JONES et al. 1985; ROMAN and GAROFF 1986; STEPHENS et al. 1986). It was also demonstrated that glycosylation of viral glycoproteins is not required for their polarized expression, or for the budding of viruses at specific membrane domains (ROTH et al. 1979; GREEN et al. 1981). Taken together, these observations indicate that the amino acid sequence of the glycoproteins determines their destination. As noted above, no obvious common structural feature is apparent in molecules destined for the same location; for example, the influenza HA and NA glycoproteins are molecules with very different three-dimensional structures and opposite transmembrane topologies, but each is targeted specifically to the apical surface of polarized epithelial cells when expressed from a cloned cDNA gene (ROTH et al. 1983a; JONES et al. 1985). Thus, the sorting signals appear to be present in structural features of the proteins which are as yet unrecognized.

Several groups have utilized recombinant DNA approaches to construct deletion mutants or chimeras between apical and basolateral proteins, in order to locate the protein domains which contain sorting signals. The cytoplasmic domains of several glycoproteins have been deleted, including the N-terminal 10 amino acids of the influenza NA glycoprotein (JONES et al. 1985), most of the C-terminal cytoplasmic domain of an envelope glycoprotein of Semliki Forest virus (ROMAN and GAROFF 1987), or the entire cytoplasmic domain of murine retrovirus gp70/p15E (KILPATRICK et al. 1986). None of these modifications affected the polarized expression of the resulting glycoprotein, indicating that the cytoplasmic domains of these proteins do not contain essential information for polarized transport.

A different result was found with the sIgA receptor, which is a bitopic membrane glycoprotein with a C-terminal cytoplasmic tail of 103 amino acids. By deletion of all but two of these amino acids, a molecule was produced with markedly altered sorting properties (MOSTOV et al. 1987). Whereas the wild-type protein is transported first to basolateral surfaces of MDCK cells and then translocated to apical surfaces (see above), the truncated protein was transported directly to apical membranes of MDCK cells. These observations indicate that a signal for basolateral expression is present in the cytoplasmic tail of this molecule. Alternatively, it is possible that the presence of a cytoplasmic tail may mask the recognition of an apical sorting signal, thus resulting in basolateral transport of wild-type molecules by default. Interaction with ligand or conformational changes may expose the apical sorting signal.

Chimeric glycoproteins containing portions of the (apical) influenza HA and the (basolateral) VSV G proteins fused together within the external domain were found to be glycosylated but blocked in transport from the RER (MCQUEEN et al. 1984). It was suggested that failure of these chimeric proteins to be expressed on the cell surface may have resulted from a transport block due to improper folding of the ectodomain. Subsequent studies have therefore focused on construction of chimeras with junctions precisely at one side or the other of the hydrophobic transmembrane domain. Replacement of the transmembrane and cytoplasmic domains of influenza HA with the corresponding region from the VSV G protein resulted in molecules which were fully transport-competent, and were expressed on the apical surfaces of epithelial cells (MCQUEEN et al. 1986; ROTH et al. 1987; COMPTON et al. 1989). These results indicated that the external domain of HA contains a signal to direct this glycoprotein to the apical membrane. The cleaved N-terminal signal sequence of the HA glycoprotein apparently does not play a role in sorting, since replacement of the HA signal with the corresponding region from the G protein resulted in a molecule that was also trasported to the apical surface (MCQUEEN et al. 1987). Another construct, in which the ectodomain of the G protein was joined to the transmembrane and cytoplasmic domains of HA, was transported to the basolateral surface (MCQUEEN et al. 1987). In contrast, when a chimera in which only the cytoplasmic domain of G was replaced with the corresponding region of HA (PUDDINGTON et al. 1987) was expressed in MDCK cell lines, were produced this glycoprotein was expressed on both the apical and basolateral surfaces. Although these two studies may appear to have generated conflicting data, the G/HA chimeric proteins differ in their transmembrane portions and in the kinetics of transport from the RER to the Golgi complex, indicating different transport properties. In addition, the distribution of proteins on surfaces of a continuous cell line may not reflect their site of insertion into plasma membranes, since the proteins could undergo redistribution during mitosis and/or cell passage.

In contrast to the above results with influenza HA and VSV G are those yielded by a chimeric molecule between murine retrovirus gp70/p15E and influenza HA. The transmembrane and cytoplasmic domains of the p15E protein were replaced with the corresponding regions from the influenza HA protein (E.B. Stephens and Compans, unpublished results). This chimeric protein was found to be expressed on the apical surface of MDCK cells. In this case, apical transport was apparently determined by the membrane anchor and/or cytoplasmic tail of HA. Thus, the available information does not present a consistent picture; in some chimeric proteins the ectodomain appears to determine the protein's destination, whereas in other constructs the membrane anchor and/or cytoplasmic domains appear to be of primary importance. These differences remain to be explained. One possibility is the existence of multiple sorting signals in different domains of some glycoproteins, one of which has a dominant effect in each construct. Alternatively, some molecules may lack any specific sorting signals, and their site of surface expression may be determined by a default pathway, as discussed above for secreted proteins.

Transmembrane domains of bitopic integral membrane proteins are very similar in structure, generally consisting of 20–24 hydrophrobic amino acids, followed by one or more positively charged residues on the cytoplasmic side of the molecule. Evidence was recently obtained, however, that such transmembrane domains possess the ability to specifically recognize one another by lateral interactions in the membrane (BORMANN et al. 1989). Thus, such interactions could play an important role in the sorting of membrane proteins, in mediating the lateral interactions which are needed to form transport vesicles containing specific sets of membrane protein.

The role of transmembrane segments in protein sorting is illustrated by recent studies on a novel group of glycoproteins that are anchored in the membrane by covalent attachment to the glycolipid glycosyl phosphatidylinositol (GPI). A variety of endogenous GPI-anchored proteins have all been localized to apical membranes of epithelial cells (LISANTI et al. 1988; see Sect. 2.2). Exogenous genes encoding GPI-anchored proteins such as decay-accelerating factor (DAF, LISANTI et al. 1989), placental alkaline phosphatase, or murine Thy-1 (BROWN et al. 1989) have been introduced into MDCK cells by transfection, and the protein products were found to be localized exclusively at the apical membranes of the transfected cells. A truncated secreted form of placental alkaline phophatase that lacks a GPI anchor was secreted nondirectionally from transfected MDCK cells (BROWN et al. 1989).

Growth hormone, which is secreted nondirectionally from MDCK cells, was found to be expressed as an apical membrane protein when engineered to contain GPI anchor sequences (LISANTI et al. 1989). HSV-1 gD is a basolateral membrane protein (SRINIVAS et al. 1986), and truncated forms of this protein are also secreted basolaterally (LISANTI et al. 1989), suggesting that signals for basolateral secretion are contained in the external domain of this molecule. However, transfer of a 37-residue-long GPI attachment signal from DAF to the ectodomain of HSV-1 gD resulted in a chimeric molecule that was targeted to the apical membranes of MDCK cells. Similarly, chimeric proteins with the ectodomain of the VSV G protein and GPI anchors from Thy-1 were also targeted to the apical membranes of MDCK cells. Together these results clearly indicate that GPI membrane anchors serve as apical targeting signals.

4.4 Polarity of Proteins on Cytoplasmic Surfaces

Recently, it was reported that the M protein of VSV, which is associated with the cytoplasmic surface of the plasma membrane, also exhibits polarity in MDCK cells and is preferentially associated with basolateral membranes (BERGMANN and FUSCO 1988). A similar distribution of M protein was found in cells infected with a glycoprotein-deficient mutant of VSV, suggesting that the localization of M protein occurred independently of the presence of viral glycoproteins on basolateral membranes. These observations indicate that proteins associated with the intracytoplasmic surfaces of cellular membranes also may possess mechanisms for recognition of specific plasma membrane domains. The cellular homolog of the Rous sarcoma virus (RSV) oncogene, $pp60^{c\text{-}src}$, is synthesized as a cytoplasmic protein which undergoes N-terminal myristylation and associates with the plasma membrane. Available evidence indicates that the membrane-bound forms of *c-src* protein are localized at the basolateral surface of polarized epithelial cells (WARREN et al. 1988). The MuLV *gag* precursor $Pr65^{gag}$ also undergoes myristylation and can form particles at the plasma membrane in the absence of envelope glycoproteins, but these glycoprotein-deficient particles appear to mature nondirectionally from both apical and basolateral membranes of transfected MDCK cells (A. Rein, A. Schultz, and R.V. Srinivas, unpublished results). Thus some, but not all, cytoplasmic membrane-associated proteins may be targeted to specific membrane domains in epithelial cells. Since such proteins are synthesized on cytoplasmic polyribosomes and do not follow the exocytic pathway in reaching the plasma membrane, the sorting mechanism for this class of proteins must differ significantly from that for membrane glycoproteins. The most obvious sorting mechanism would involve direct recognition of apical or basolateral plasma membrane components by such proteins. Since the lipid compositions of apical and basolateral membranes appear to be indistinguishable on their inner leaflets (VAN MEER and SIMONS 1986), it is more likely that cellular membrane-associated proteins are involved as recognition sites.

5 Possible Mechanisms for Sorting Apical Vs Basolateral Membrane Proteins

5.1 Default Pathways

Sorting of apical and basolateral membrane proteins could be achieved even if only one class of proteins, either apical or basolateral, contained specific sorting signals. These signals could then direct the incorporation of the proteins into a specific population of transport vesicles which would deliver the proteins to one plasma membrane domain. Other proteins which lack these sorting signals would be excluded from such vesicles, packaged into a different set of vesicles, and delivered to the other membrane domain by bulk flow, thus being sorted in the absence of a specific signal by a unidirectional default pathway. In the case of secreted proteins, evidence has been obtained in support of such unidirectional pathways in hepatocytes and intestinal epithelial cells, and for a bidirectional pathway in MDCK cells, as discussed above (Sect. 3.1). However, the nature of default pathways for membrane proteins in polarized epithelial cells is not known. It has been proposed that the default pathway is directed to apical membranes (PFEFFER and ROTHMAN 1987), to basolateral membranes (BARTLES and HUBBARD 1988), or is nondirectional (MATLIN 1986b). Since recent results indicate that cell-type-specific differences exist in the default pathway for secreted proteins, it is possible that such differences will also be observed for membrane proteins. It may be possible to obtain information about the default pathway by analyzing the site of expression of membrane proteins which are not normally found on the surfaces of polarized epithelial cells. Studies of default pathways will provide useful insights into the sorting of endogenous proteins, and molecules transported by default will be useful for experimental studies. Identification of putative sorting signals would be facilitated by substitution of candidate sequences from sorted proteins into proteins destined for transport by default.

Some evidence suggests that a basolateral default pathway may exist, at least in some cell type; (a) Studies with hepatocytes and intestinal epithelial cells have indicated that several apical membrane proteins first arrive at the basolateral membranes prior to their delivery to apical membranes (BARTLES et al. 1987; HAURI et al. 1979; QUARONI et al. 1979a, b). (b) Immature intestinal epithelial cells express basolateral proteins normally whereas transport of apical proteins is abortive (LOUVARD et al. 1985). (c) Epithelial cells have a functional basolateral-to-apical transcytotic pathway which can redistribute some proteins to apical membranes from the site of their initial delivery at basolateral membranes. Also, in MDCK cells, the apical proteins are stored in a novel vacuolar apical storage compartment when maintained in calcium-free medium (VEGA-SALAS et al. 1987, 1988). Thus, the apical surfaces of the epithelia may resemble a specialized organelle formed during development and reorganization of epithelia, while the basolateral membranes may resemble the plasma membrane of nonpolarized cells.

5.2 The Interaction of Sorting Signals with Cellular Proteins Involved in Vesicular Transport

As described above, viral proteins destined for distinct plasma membrane domains of epithelial cells are localized in the same region of the Golgi complex, but appear to exit from trans-Golgi cisternae into distinct populations of vesicles which could mediate their transport to the plasma membrane domains. It is likely, therefore, that the critical event in the process of protein sorting occurs at the stage of formation of such transport vesicles, and that the sorting signals in membrane proteins are recognized at this stage of the transport process. The sorting of membrane proteins may occur by the segregation of proteins into microdomains by lateral movement in the plane of the membrane, followed by the formation of a vesicle from such a patch of membrane by budding and pinching off. The simplest model for sorting of membrane proteins into distinct populations of transport vesicles would involve their self-association into clusters by lateral interactions among the proteins themselves, followed by budding to form a transport vesicle. The sorting signals would then be responsible for the subsequent recognition between these transport vesicles and the target membrane. Such a model would require that newly synthesized plasma membrane proteins should be present in the Golgi complex in concentrations sufficiently high to mediate self-association. While this could possibly be the case with the membrane glycoproteins of some enveloped viruses, which are synthesized in large quantities, it is unlikely to be the case with most cellular membrane proteins. In addition, such a model would predict that an essential recognition event would be mediated by interactions between the target membrane and the cytoplasmic domains of the membrane glycoproteins, which are exposed on the external surfaces of transport vesicles. However, as already noted, the cytoplasmic domains of several glycoproteins do not contain information essential for their polarized expression. We therefore consider such a self-sorting process to be unlikely.

It is more likely that the sorting signals in plasma membrane proteins are recognized by lateral interactions with components of trans-Golgi membranes, and that this interaction leads to the clustering of membrane proteins into microdomains. SIMONS and FULLER (1985) have pointed out that the clustering of membrane proteins and budding of a vesicle during vesicular transport is similar in many respects to the final step in assembly and release of enveloped viruses by budding at the cell surface. However, the viral proteins exhibit opposite orientations with respect to the curvature of the membrane during these two processes. During vesicular transport, the large external domains of viral glycoproteins remain sequestered in the interior of transport vesicles. In contrast, during the budding process involved in virus assembly, these same domains are

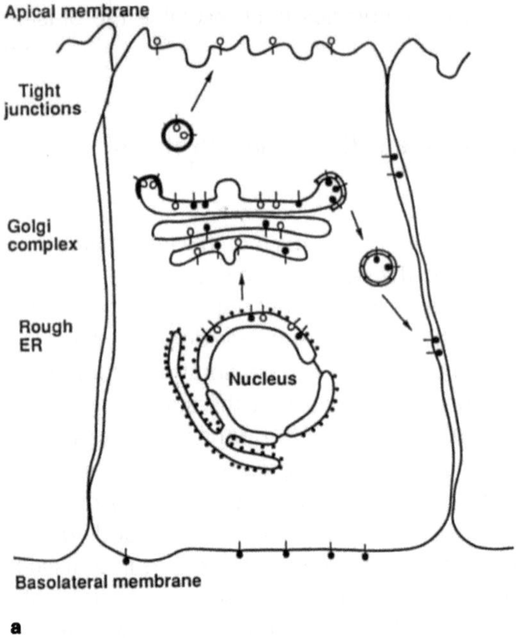

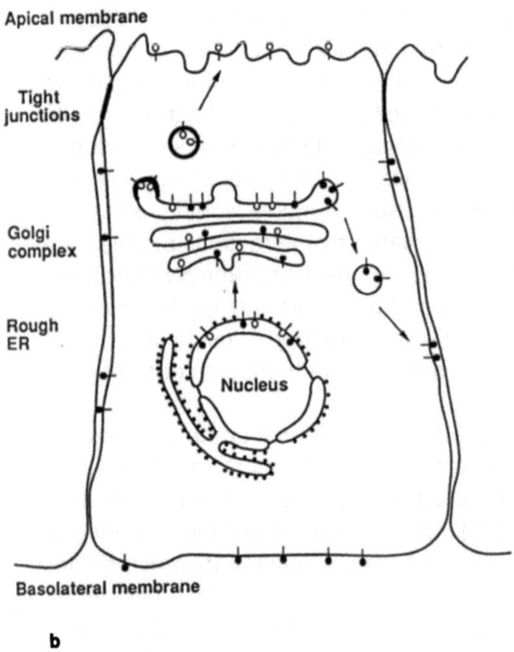

Fig. 2 a–c. Alternative models for the sorting of membrane glycoproteins in polarized epithelial cells. **a** Polarized delivery of proteins to the apical and basolateral membranes from the trans-Golgi network in distinct transport vesicles. Specific sorting signals on both proteins mediate their association with distinct sets of cellular molecules (*thick lines* and *open double lines*) to form distinct transport vesicles, which deliver the proteins directly to the correct destination. **b** Specific sorting signals on apical proteins mediate their inclusion into one set of transport vesicles (*thick lines*) at the stage of exit from the trans-Golgi network, while basolateral proteins are packaged into a different set of vesicles by default. The apical and basolateral proteins are delivered directly to appropriate target membranes from the trans-Golgi network. **c** Apical as well as basolateral proteins are first delivered to the basolateral membranes, followed by selective endocytosis of apical proteins into distinct vesicle populations (*thick lines*), which transport them to the apical membranes

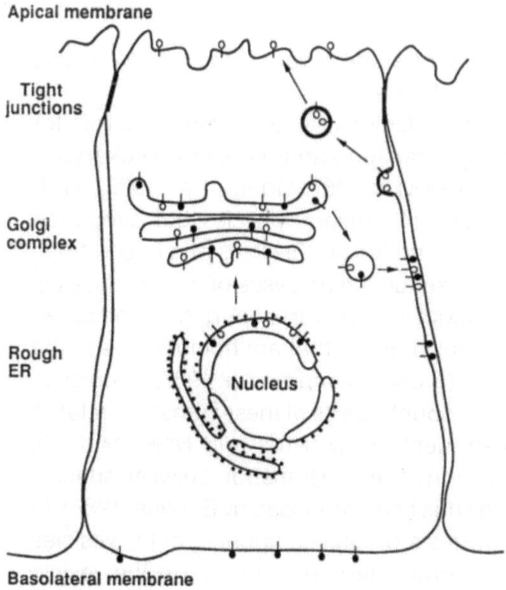

c

located on the external surface of the budding virions in a structure with the opposite curvature. The presence of cellular proteins in the transport vesicles, and their absence in budding virions, may account for the differences in packing which would be expected if the same protein is to be incorporated into budding structures with such opposite curvatures.

Schematic representations of three different protein sorting scenarios are shown in Fig. 2. In each model, sorting is thought to occur at the exit from the Golgi complex. In Fig. 2a, both apical and basolateral proteins are positively sorted into distinct vesicle populations. Proteins that lack any sorting signals would be delivered randomly to both apical and basolateral membranes. In Fig. 2b, apical proteins are positively sorted into one set of vesicles, while basolateral proteins are packaged into a different set of vesicles by default. In Fig. 2c, apical and basolateral membranes are both delivered to the basolateral membranes, and the apical proteins are sorted thereafter by an endocytic pathway and transported to apical membranes.

Subsequent to the sorting of membrane proteins into distinct sets of transport vesicles, the vesicles must be delivered to the appropriate target membrane. The molecular basis of this interaction is unknown, but it presumably result from specific interactions between components on the external surface of the transport vesicle and those on the cytoplasmic surface of the target membrane. Since deletion of the cytoplasmic domains did not prevent the directional transport of several viral membrane proteins, it is likely that the cellular proteins which mediate sorting at the stage of vesicle formation are also involved in recognition of the target membrane.

Biochemical and genetic evidence has demonstrated an important role for GTP hydrolysis in the regulation of protein transport and secretion in eukaryotic cells (NOVICK et al. 1980; SALMINEN and NOVICK 1987; GOUD et al. 1988). GTP hydrolysis is not required for the formation of transport vesicles, but is required for subsequent steps involving uncoating and fusion (BECKERS and BALCH 1989; BALCH et al. 1984a, b). Gene cloning and sequence analyses of the yeast genes SEC4 and YPT-1, which are associated with vesicular transport, have indicated that these genes encode small (~23 kDa) proteins that are homologous to the ras superfamily of GTP-binding proteins (SALMINEN and NOVICK 1987; GOUD et al. 1988; MELANCON et al. 1987). Mammalian counterparts of these transport-related GTP-binding proteins have also been identified (SANTOS and NEBRADA 1989; ZAHRAOUI et al. 1989). A model for mediation of vesicle transport between specific compartments by GTP-binding proteins has been proposed by BOURNE (1988). In this model, a transport GTPase recognizes a protein component of the vesicles emerging from a donor membrane to form a ternary complex on the vesicle surface. This ternary complex interacts specifically with a component on the acceptor membrane and catalyzes GTP hydrolysis, which in turn leads to initiation of the fusion process resulting in delivery of vesicle contents to the target acceptor membrane. Available evidence indicates that distinct GTP-binding proteins may be involved in different stages of membrane traffic. For example, yeast YPT-1 (and mammalian rab) protein may be involved in vesicular transport between the ER and the Golgi complex, while the yeast SEC4 protein is thought to play a role in vesicular transport between the trans-Golgi network and the plasma membrane (GOUD et al. 1988; SCHMITT et al. 1988; SEGEV et al. 1988; ZAHRAOUI et al. 1989; BALCH 1989). Identification of multiple small GTP-binding proteins in a variety of mammalian cells makes it plausible that one set of these GTP-binding proteins may be involved in transport of apical proteins, while a different subset may be involved in transport of basolateral proteins. Recently, several in vitro systems have been described in which the transport of proteins between the Golgi complex and the plasma membrane has been demonstrated (BALCH et al. 1984a, b; WOODMAN and EDWARDSON 1986; BECKERS et al. 1987; SIMONS and VIRTA 1987; BENNETT et al. 1988). Such systems will facilitate the isolation of transport vesicles and the identification of both the cellular molecules involved in the formation of transport vesicles and the nature of the recognition event at the target membrane.

5.3 Role of Lipids in Protein Sorting

Studies on the composition and transport of plasma membrane lipids in polarized epithelial cells have suggested their possible involvement in protein targeting. The apical plasma membranes of epithelial cells show a two- to fourfold enrichment in glycosphingolipids, while the basolateral membranes show a two- to fourfold enrichment in phosphatidylcholine (PC) (BRASITUS and SCHACHTER 1980; MEIER et al. 1984; MOLITORIS and SIMON 1985, 1986). Since tight junctions form a barrier for the diffusion of lipids between the exoplasmic leaflet, but not the inner leaflet, of apical and basolateral membranes (DRAGSTEN et al. 1981, 1982), the inner leaflets of the two domains are thought to be indistinguishable in lipid composition (VAN MEER and SIMONS 1988). This suggested that nearly all the glycosphingolipids are located at the exoplasmic leaflet of the apical membrane, while phospholipids are mainly localized in the cytoplasmic leaflet.

Previous studies have indicated that transport of lipids from the Golgi complex to the plasma membranes involves vesicular transport. Using a fluorescent ceramide analog, VAN MEER et al. (1987) observed that upon incubation of labeled liposomes with MDCK cells, the fluorescent probes partition into membranes, and are trapped in the Golgi complex. The fluorescent probes are then incorporated into the luminal leaflet of the Golgi complex as newly synthesized C6-NBD-sphingomyelin and C6-NBD-glucosyl ceramide, and are transported to the plasma membrane in vesicles. Furthermore, the cell surface delivery was found to be polar, in that per unit surface area 10 times more C6-NBD-glucosyl ceramide and 3 times more C6-NBD-sphingomyelin were delivered to the apical plasma membrane domain. Based on these observations, VAN MEER and SIMONS (1988) have proposed a vesicular transport model to explain sorting of lipids. In this model, it is assumed that phospholipid asymmetry between the two inner and outer leaflets of the bilayer is established in the endoplasmic reticulum and the asymmetric distribution of choline-containing and amino-containing phospholipids is maintained during the vesicular transport processes between the ER and the cell surface. Subsequently, sphingolipids synthesized in the Golgi complex are incorporated preferentially into the exoplasmic leaflet. The specific event in lipid sorting would then involve generation of sphingolipid microdomains in the luminal leaflet of the trans-Golgi network, which would then bud out as a vesicle destined for apical membranes. Because of their sphingosine backbones, sphingolipids have a tendency to associate by hydrogen bond formation, and form separate domains in model membrane systems (PASCHER 1976; THOMPSON and TILLACK 1985; THOMPSON et al. 1986). Basolaterally destined vesicles would be derived from sphingolipid-deficient (and consequently PC-enriched) regions of the trans-Golgi network.

Since the delivery of apical lipids and apical proteins to the plasma membrane in MDCK cells show the same kinetics and both are slowed down by a 20 °C temperature block (SIMONS and VIRTA 1987; VAN MEER and SIMONS 1988), it was suggested that the two processes may be related, and that the sorting of plasma membrane proteins in epithelial cells may involve cooperativity between lipids and proteins (VAN MEER and SIMONS 1988). For example, sphingolipids could bind to a putative sorting signal on apical proteins and concentrate them in a microdomain which would bud to form an apically destined vesicle. The recent finding that the GPI anchor represents an apical sorting signal provides a possible example where association among lipids may lead to apical transport.

Recently it was reported that the cleaved fusion (F) glycoprotein of parainfluenza viruses and the influenza HA glycoprotein at ~pH 5.5 have lipid transport activity (DEMEL et al. 1987). Interestingly, these glycoproteins also mediate membrane fusion. A search for sequence homology using Seller's TT algorithm has revealed a region of alignment between the domain of cellular PC-specific transport protein from beef liver, which interacts with fatty acids, and the segment of HA which forms a long helix along the HA spike extending out from the lipid bilayer. Also, the F glycoprotein of Sendai virus, another paramyxovirus, has been shown to be capable of binding cholesterol (ASANO and ASANO 1988), which is enriched in the apical membranes of epithelial cell. All the above-mentioned viral glycoproteins are sorted to the apical domains of epithelial cells. It is possible that lipid-binding domains might have a common evolutionary origin, and some sequences might have evolved for binding PC and others for sphingolipids. Although the homology of HA and parainfluenza virus F proteins was found with transport proteins specific for PC (which is enriched in the basolateral membrane of epithelial cells), the observed homology may merely be indicative of lipid-binding abilities, and not a specificity for PC. Further studies on the ability of various apical and basolateral proteins to interact with different lipid molecules should provide insight into the possible role of lipids in protein sorting.

5.4 Role of Protein Oligomerization in Transport and Sorting

It is now well established that a number of membrane proteins are oligomeric structures. The formation of oligomers appears to be a requirement for movement through the earliest stage of the transport pathway, namely exit from the RER (COPELAND et al. 1986; GETHING et al. 1986). Mutants which are blocked in oligomerization are blocked in transport from the RER (GETHING et al. 1986; COPELAND et al. 1988). In some cases, such proteins appear to be bound to proteins which are permanent residents of the RER, although it is not clear whether this is the basis for their retention in ER.

The oligomerization of membrane proteins may be a requirement for the function of sorting signals in some membrane glycoproteins. In particular, the lateral clustering of proteins into microdomains within the Golgi complex may involve the formation of identical protein-protein bonds, which would be facilitated if the proteins involved are oligomers. In contrast, such protein-protein interactions would not be a requirement for transport of a membrane protein to the cell surface via a default pathway or for lipid-mediated protein sorting. The essential feature for transport by default would presumably be the lack of a specific sorting or retention signal; the protein must not interact stably with other molecules that are permanent residents of one of the compartments of the transport pathway. Oligomeric or monomeric proteins which lack specific sorting or retention signals may be carried to the cell surface via the same default pathway.

In contrast to the results obtained with certain membrane glycoproteins, a similar requirement for oligomerization has not been reported for transport of secretory proteins to the cell surface. If the sorting of such proteins is mediated by binding to membrane-bound receptor proteins, a process similar to that involving the receptor for mannose-6-phosphate in the sorting of lysosomal proteins, it would not be expected that oligomerization is required for such interactions.

6 Concluding Remarks

Despite the recent improvement in our understanding of the processes involved in establishment and maintenance of epithelial membrane polarity, several key issues remain unsettled. A central event in protein sorting is the vesicular transport process, and unresolved questions include (a) how specific transport vesicles are formed, and (b) how the vesicles are directed to specific target membranes. In Sect. 5 we have considered several possibilities, which remain to be distinguished by experimental evidence. This will require the identification of putative cellular "sorting proteins". Evidence that such molecules may exist for secreted proteins has been obtained with the recent finding that a 25-kDa family of proteins bind to different secretory proteins along the regulated pathway (CHUNG et al. 1989). Preliminary evidence (CHUNG et al. 1989) indicated that this class of proteins is immunologically related among different species; also, differences in the affinity of various secretory granule proteins to the insulin-binding transport protein suggested that subspecies of this molecule with varying specificities may exist. This raises the interesting possibility that such proteins may belong to a family of closely related genes, and it is not unlikely that the proteins involved in sorting of apical and basolateral belong to such a family. In this regard, molecular characterization of the proteins involved in the sorting of secretory proteins should provide valuable information. Other cellular proteins that may be involved in protein sorting include members of the ras superfamily of GTP-binding proteins. Low-molecular-weight GTP-binding proteins have been identified in highly purified transport vesicles emerging from the trans-Golgi network in baby hamster kidney cells (K. Simons, cited in BALCH 1989). Whether such proteins are also present in transport vesicles from different epithelial cells, and whether distinct GTP-binding proteins are associated with vesicles carrying apical or basolateral proteins, remains to be established.

A second aspect of this problem concerns the putative "sorting signals" or the structural features on the proteins which determine the destination of the molecule. Despite numerous studies involving genetically engineered proteins, and molecular chimeras between apical and basolateral protein, no consensus has emerged on either the nature of the sorting signals or their location (e.g., external, membrane-spanning, or cytoplasmic domains) on the transmembrane molecules. One limitation in these experiments is our inability to assess the effect of any modifications on the conformation and folding of the proteins, which may mask existing sorting signals or expose cryptic sorting signals. The possibility exists that the sorting signals are degenerate, like the signals that determine protein translocation across the membranes of the ER or mitochondria (VON HEIJU 1985; ROISE et al. 1986). Nevertheless, additional studies of chimeric proteins with appropriate checks on protein conformation should be useful.

Once the specific transport vesicles are formed, they need to be targeted to a specific membrane compartment, and no information is available on how this is accomplished. A valuable approach in identifying components of cellular machinery has involved isolation and characterization of cellular mutants defective in a particular function. This approach, although most effectively employed in bacteria and yeast, has also been used successfully with higher eukaryotes. Several MDCK mutants have been described, including a mutant that is apparently defective in transport of a basolateral viral glycoprotein but not in that of an apical virus (MAYNE and COMPANS 1988). Such MDCK cell mutants with defects in specific stages of the transport pathway may provide a useful tool to study cellular components involved in protein sorting. In recent years, methods have been described which facilitate isolation of transport vesicles carrying apical and basolateral proteins from mechanically perforated MDCK cells (BENNETT et al. 1988). This should facilitate further studies of apical and basolateral transport vesicles, and identification of sorting components. Another approach which may prove useful involves transfer of genes into nonepithelial cells to identify properties specific to epithelia (MEGE et al. 1988; FRIEDERICH et al. 1989).

References

Almers W, Stirling C (1984) Distribution of transport proteins over animal cell membranes. J Membr Biol 77: 169–186

Alonso FV, Compans RW (1981) Differential effect of monensin on enveloped viruses that form at distinct plasma membrane domains. J Cell Biol 89: 700–705

Alonso-Caplen FV, Compans RW (1983) Modulation of glycosylation and transport of viral membrane glycoproteins by a sodium ionophore. J Cell Biol 97: 659–668

Alton JF, Brown CDA, Ogden P, Simmons NL (1982) K+ transport in tight epithelial monolayers of MDCK cells. J Membr Biol 65: 99–109

Asano A, Asano K (1988) Binding of cholesterol and inhibitory peptide derivatives with fusogenic hydrophobic sequence of F glycoprotein of HVJ (Sendai virus): possible implication in the fusion process. Biochemistry 27: 1321–13

Balch WE (1989) Biochemistry of interorganellar transport. A new frontier in enzymology emerges from versatile in vitro model systems. J. Biol. Chem. 264: 16965–16968

Balch WE, Dunphy WG, Braell WA, Rothman JE (1984a) Reconstitution of transport of protein between successive ompartments of the Golgi measured by the coupled incorporation of N-acetyl glucosamine. Cell 39: 405–416

Balch WE, Glick BS, Rothman JE (1984b) Sequential intermediates in the pathway of intercompartmental transport in a cell free system. Cell 39: 525–536

Baron R, Neff L, Brown W, Courtoy PJ, Louvard D, Farquhar MG (1988) Polarized secretion of lysosomal enzymes: codistribution of cation independent mannose-6-phosphate receptors and lysosomal enzymes along the osteoclast exocytic pathway. J Cell Biol 106: 1863–1872

Barth A, Schultz H, Neubert K (1974) Untersuchungen zur Reinigung und Characterisierung der Dipeptidylaminopeptidase IV. Acta Biol Med Ger 32: 157–174

Bartles JR, Hubbard AL (1988) Plasma membrane protein sorting in epithelial cells: do secretory pathways hold the key? Trend Biochem Sci 13: 181–184

Bartles, JR, Feracci HM, Stieger B, Hubbard AL (1987) Biogenesis of the rat hepatocyte plasma membrane in vivo: comparison of the pathways taken by apical and basolateral proteins using subcellular fractionation. J Cell Biol 105: 1241–1251

Beckers CJM, Balch WE (1989) Calcium and GTP: essential components in vesicular trafficking between the endoplasmic reticulum and Golgi apparatus. J. Cell Biol 108: 1245–1256

Beckers CJM, Keller DS, Balch WE (1987) Semiintact cells permeable to micromolecules: use in reconstitution of protein transport from endoplasmic reticulum to Golgi complex. Cell 50: 523–534

Behrens J, Birchmeier W, Goodman SL, Imhoff BA (1985) Dissociation of Madin-Darby canine kidney epithelial cells by monoclonal antibody anti-Arc-1: mechanistic aspects and identification of the antigen as a component related to uvomorulin. J Cell Biol 101: 1307–1315

Bennet G, Carlet G, Wild G, Parsons S (1984) Influence of colchicine and vinblastine on the intracellular migration of membrane glycoproteins. III. Inhibition of intracellular migration of membrane glycoproteins in rat columnar cells and hepatocytes as visualized by light and electron microscope radioautography after 3H-fucose injection. Am J Anat 170: 545–566

Bennett MK, Wandinger-Ness A, Simons K (1988) Release of putative exocytic transport vesicles from perforated MDCK cells. EMBO J 7: 4075–4085

Bergmann JE, Fusco PJ (1988) The M protein of vesicular stomatitis virus associates specifically with the basolateral membranes of polarized epithelial cells independently of the G protein. J Cell Biol 107: 1707–1715

Blok AJ, Ginsel LA, Mulder-Stapel AA, Onderwater JUM, Daems W (1981) The effect of colchicine on the intracellular transport of 3H-fucose labelled glycoproteins in the absorptive cells of cultured human small-intestinal issue. Exp Cell Res 215: 1–12

Blok J, Air GM, Laver WG, Ward CW, Lilley GG, Wood EF, Roxburgh CM, Inglis AS (1982) Studies on the size, chemical composition, and partial sequence of th neuraminidase (NA) from type A influenza viruses show that the N-terminal region of NA is not processed, but serves to anchor the NA in the viral membrane. Virology 119: 109–121

Boller K, Vestweber D, Kemler R (1985) Cell adhesion molecule uvomorulin is localized in the intermediate junctions of adult intestinal epithelial cells. J Cell Biol 100: 327–332

Borgers M (1973) The cytochemical application of new potent inhibitors of alkaline phospatases. J Histochem Cytochem 21: 812–824

Bormann BJ, Knowles WJ, Marchesi VT. (1989) Synthetic peptides mimic assembly of transmembrane glycoproteins. J Biol Chem 264: 4033–4037

Bourne H (1988) Do GTPases direct membrane traffic in secretion? Cell 53: 669–671

Brasitus TA, Schachter D (1980) Lipid dynamics and lipid-protein interactions in rat enterocyte basolateral and microvillus membranes. Biochemistry 19: 2763–2769

Brown DA, Crise B, Rose JK (1989) Mechanism of membrane anchoring affects polarized expression of two proteins in MDCK cells. Science 245: 1499–501

Brown WJ, Constantinescu E, Farquhar MG (1984) Redistribution of mannose-6-phosphate receptors induced by tunicamycin and chloroquine. J Cell Biol 99: 320–326

Burgess TL, Kelly RB (1987) Constitutive and regulated secretion of proteins. Ann Rev Cell Biol 3: 243–293

Caplan MJ, Anderson HC, Palade GE, Jamieson JD (1986) Intracellular sorting and polarized cell surface delivery of (Na+, K+)ATPase, an endogenous component of MDCK cell basolateral plasma membranes. Cell 46: 623–631

Caplan MJ, Stow JL, Newman AP, Madri J, Anderson HC, Farquhar MG, Palade GE, Jamieson JD (1987) Dependence on pH of polarized sorting of secreted proteins. Nature 329: 632–635

Carlison RW, Wada HG, Sussman HH (1976) The plasma membranes of human placenta; isolation and characterization of protein and glycoprotein subunits. J Biol Chem 251: 4139–4146

Casanova JE, Breitfeld PP, Ross SA, Mostov KE (1990) Phosphorylation of the polymeric immunoglobulin receptor is required for its efficient transcytosis. Science 248: 742–745

Cereijido M, Stefani E, Martinez-Palomo A (1980) Occluding junctions in a cultured transporting epithelium: structural and functional heterogeneity. J Membr Biol 53: 19–32

Chambard M, Gabrion J, Mauchamp J (1981) Influence of collagen gel on the orientation of epithelial cell polarity: follicle formation from isolated thyroid cells and from preformed monolayers. J Cell Biol 91: 157–166
Chung K, Walter P, Aponte GW, Moore H-H (1989) Molecular sorting in the secretory pathway. Science 243: 192–197
Claude P (1978) Morphological factors influencing transepithelial permeability: a model for the resistance of zonula occludens. J Membr Biol 39: 219–232
Claude P, Goodenough DA (1973) Fracture faces of zonulae occludentes from "tight' and "leaky" epithelia. J Cell Biol 58: 390–400
Colas B, Maroux S (1980) Simultaneous isolation of brush border and basolateral membranes from rabbit entrocytes: presence of brush border hydrolases in the basolateral membrane of rabbit enterocytes. Biochim Biophys Acta 600: 406–420
Compans RW, Klenk H-D (1979) Viral membranes. Compr Virol 13: 293–407
Compans RW, Klenk H-D, Caliguiri LA, Choppin PW (1970) Influenza virus proteins. I. Analysis of polypeptides of the virion and identification of spike glycoproteins. Virology 42: 880–889
Compans RW, Roth MG, Alonso FV, Srinivas RV, Herrler G, Melsen LR. (1982) Do viral maturation sites influence disease processes? In: Mackenzie JS (ed) Viral diseases in South-East Asia and the Western Pacific. Academic, New York, pp 328–332
Compton T, Ivanov IE, Gottlieb T, Rindler M, Adesnik M, Sabatini DD (1989). A sorting signal for the basolateral delivery of the vesicular stomatitis virus (VSV) G protein lies in its luminal domain: analysis of the targeting of VSV G-influenza hemagglutinin chimeras. Proc Natl Acad Sci 86: 4112–4116
Copeland CS, Doms RW, Bolzau EM, Webster RG, Helenius A (1986) Assembly of influenza hemagglutinin trimers and its role in intracellular transport. J Cell Biol 103: 1179–1191
Copeland CS, Zimmer K-P, Healey GA, Mellman I, Hellenius A (1988) Folding, trimerization, and transport are sequential events in the biogenesis of influenza virus hemagglutinin. Cell 53: 197–209
Cross GAM (1987) Eukaryotic protein modification and membane attachment via phosphatidyl inositol. Cell 179-181
Danielsen EM, Vyas JP, Kenny AJ (1980) A neutral endopeptidase in the microvillar membrane of pig intestine. Biochem J 191: 645–648
Demel RA, Sehgal PB, Landsberger FR (1987) A structural model for fusion of viral and cellular membranes. In: Mahy BWJ, Kolakofsky D (eds) The biology of negative strand viruses. Elsevier, Amsterdam, pp 26–32
Desnuelle P (1979) Intestinal and renal aminopeptidase: a model of a transmembrane protein. Eur J Biochem 101: 1–11
Douglas AP, Kerley R, Isselbacher KJ (1972) Preparation and characterization of the lateral and basal plasma membranes of the rat intestinal epithelial cell. Biochem J 128: 1329–1338
Dragsten PR, Blumenthal R, Handler JS (1981) Membrane asymmetry in epithelial: is the tight junction a barrier to diffusion in the plasma membrane? Nature 294: 718–722
Dragsten PR, Handler JS, Blumenthal R (1982) Fluorescent membrane probes and the mechanism of maintenance of cellular asymmetry in epithelia. Fed Proc 41: 48–53
Dubois-Dalcq M, Holmes K, Rentier B (1984) Assembly of enveloped RNA viruses. Springer, Vienna
Ducibella T, Albertini DF, Anderson E, Biggers JD (1975) The preimplantation embryo: characterization of intercellular junctions and their appearance during development. Dev Biol 47: 45–58
Eilers U, Klumperman J, Hauri H-P (1989) Nocadozole, a microtubule-active drug, interferes with apical protein delivery in cultured intestinal epithelial cells (CaCo-2). J Cell Biol 108: 13–22
Ellinger A, Pavelka M, Gangl A (1983) Effect of colchicine on rat small intestinal absorptive cells. II. Distribution of label after incorporation of {3H} fucose into plasma membrane glycoproteins. J Ultrastruct Res 85: 260–271
Evans WH (1980) A biochemical dissection of the functional polarity of the plasma membrane of the hepatocyte. Biochim Biophys Acta 604: 27–64
Farquhar MG, Palade GE (1963) Junctional complexes in various epithelia. J Cell Biol 17: 375–412
Fergusson MAJ, Williams AF (1988) Cell surface anchoring of proteins via glycosyl phosphatidyl inositol structures. Ann Rev Biochem 57: 285–320
Fey EG, Wan KM, Penman S (1984) Epithelial cytoskeletal framework and nuclear matrix-intermediate filament scaffold: three dimensional organization and protein composition. J Cell Biol 98: 1973–1984
Fields S, Winter G, Brownlee GG (1981) Structure of the neuraminidase gene in influenza virus A/PR/8/34. Nature 290: 213–217

Forstner GG, Tanaka K, Isselbacher KJ (1968) Lipid composition of the isolated rat intestinal microvillus membrane. Biochem J 109: 51–59

Friederich E, Huet C, Arpin M, Louvard D (1989) Villin induces microvilli growth and actin redistribution in transfected fibroblasts. Cell 59: 461–75

Fuller SD, Bravo R, Simons K (1985) An enzymatic assay reveals that proteins destined for the apical and basolateral domains of an epithelial cell line share the same late Golgi comaprtments. EMBO J 4: 297–307

Fuller SD, Simons K (1986) Transferrin receptor polarity and recycling accuracy in "tight" and "leaky" strains of Madin-Darby canine kidney cells. J Cell Biol 103: 1767–1779

Garoff H, Kondor-Koch C, Riedel H (1982) Structure and assembly of alpha viruses. In: Heple W et al. (eds) Current topics in microbiology and immunology, vol 99. Springer, Berlin Heidelberg New York, pp 1–50

Garoff H, Frischauf A-M, Simons K, Lehrach H, Delius H (1980) Nucleotide sequence of cDNA coding for Semliki Forest Virus membrane glycoproteins. Nature 288: 236–241

George SG, Kenny AJ (1973) Studies on the enzymology of purified preparations of the brush border from rabbit kidney cortex. Exp Cell Res 51: 123–140

Gething M-J, Sambrook J (1982) Construction of influenza hemagglutinin genes that code for intracellular and secreted forms of the protein. Nature 300: 598–603

Gething M-J, McCammon K, Sambrook J (1986) Expression of wild-type and mutant forms of influenza hemagglutinin: The role of folding in intracellular transport. Cell 46: 939–950

Gipson IK, Grill Sm, Spurr SJ, Brennan SJ (1983) Hemidesmosome formation in vitro. J Cell Biol 97: 849–857

Gomori G (1941) The distribution of phosphatase in normal organs and tissues. J Cell Comp Physiol 17: 71–83

Gonzalez A, Rizzolo L, Rindler M, Adesnik M, Sabatini DD (1987) Nonpolarized secretion of the truncated forms of the influenza hemagglutinin and the vesicular stomatitis G protein from the polarized epithelial cells. Proc Natl Acad Sci USA 84: 3738–3742

Gottlieb TA, Gonzalez A, Rizzolo L, Rindler MJ, Adesnik M, Sabatini DD (1986a) Sorting and endocytosis of viral glycoproteins in transfected polarized epithelial cells. J Cell Biol 102: 1242–1255

Gottlieb TA, Beaudry G, Rizzolo L, Colman A, Rindler MJ, Adesnik M, Sabatini DD (1986b) Secretion of endogenous and exogenous proteins from polarized MDCK cell monolayers. Proc Natl Acad Sci USA 83: 2100–2104

Goud B, Salminen A, Walworth NC, Novick PJ (1988) A GTP-binding protein required for secretion rapidly associates with secretory vesicles and the plasma membrane in yeast. Cell 53: 753–768

Green RF, Meiss HK, Rodriguez-Boulan EJ (1981) Glycosylation does not determine segregation of viral envelope proteins in plasma membranes of epithelial cells. J Cell Biol 89: 230–239

Griepp EB, Dolan WJ, Robbins ES, Sabatini DD (1983) Participation of plasma membrane proteins in the formation of tight junctions by cultured epithelial cells. J Cell Biol 96: 693–702

Griffiths G, Simons K (1986) The trans Golgi network: sorting at the exit site of the Golgi complex. Science 234: 438–443

Hannafin J, Kinne-Saffran E, Friedman D, Kinne R (1983) Presence of a sodium-potassium cotransport system in the rectal gland of *Squalus acanthias*. J membr Biol 75: 73–83

Hashimoto S, Fumagalli G, Zanini A, Meldolesi J (1987) Sorting of three secretary proteins to distinct secretory granules in acidophilic cells of cow anterior pituitary. J Cell Biol 105: 1579–1586

Hauri H-P, Quaroni A, Isselbacher J (1979) Biosynthesis of intestinal plasma membrane: post-translational route and cleavage of sucrose-isomaltase. Proc Natl Acad Sci USA 85: 1942–1946

Heidrich H-G, Kinne R, Kinne-Saffron E, Hannig K (1972) The polarity of proximal tubule cell in rat kidney: different surface charges for the brush border microvilli and plasma membranes from the basal infolding. J Cell Biol 54: 232–245

Hooper NM, Low MG, Turner AJ (1987) Renal dipeptidase is one of the membrane proteins released by PI-specific PLC. Biochem J 244: 465–469

Hoope CA, Connolly TP, Hubbard AL (1985) Transcellular transport of polymeric IgA in the rat hepatocyte: biochemical and morphological characterization of the transport pathway. J Cell Biol 101: 2113–2123

Hull BE, Staehlin LA (1979) The terminal web: a reevaluation of its structure and function. J Cell Biol 81: 67–82

Ikezawa H, Yamanegi M, Taguchi R, Miyashita I, Ohyabu T (1976) A phosphatidylinositiol phospholipase C (Phosphatasemia factor) of *Bacillus cereus*. Biochim Biophys Acta 450: 154–164

Imhoff BA, Vollmers P, Goodman SL, Birchmeier W (1983) Cell–cell interaction and polarity of epithelial cells: specific peturbation using a monoclonal antibody. Cell 35: 667–675
Inoue M, Kinne R, Tran T, Blempica L, Arias IM (1983) Rat liver canalicular membrane vesicles: isolation and topological characterization. J Biol Chem 258: 5183–5188
Jaffe LF (1981) The role of ionic currents in establishing developmental pattern. Philos Trans R Soc London [Biol] 295: 553–566
Jones LV, Compans RW, Davis AR, Bos TJ, Nayak DP (1985) Surface expression of influenza virus neuraminidase, an aminoterminally anchored viral membrane glycoprotein, in polarized epithelial cells. Mol Cell Biol 5: 2181–2189
Kachar B, Reese TS (1982) Evidence for the lipidic nature of tight junction strands. Nature (London) 227: 680–685
Kashgarian M, Biemesderfer D, Caplan M, Forbush B III (1985) Monoclonal antibody to Na, K-ATPase: immunocytochemical localization along nephron segments. Kidney Int 29: 10–20
Kawai K, Fujita M, Nakao M (1974) Lipid components of two different regions of an intestinal epithelial cell membrane of mouse. Biochim Biophys Acta 369: 222–233
Kenny AJ, Maroux S (1982) Topology of microvillar membrane hydrolases of kidney and intestine. Physiol Rev 62: 91–128
Keonig B, Ricapito S, Kinne R (1983) Chloride transport in thick ascending limb of Henle's loop: potassium dependence and stoichiometry of the NaCl cotransport system in plasma membrane vesicles. Pflugers Arch 399: 173–179
Kerjaschki D, Noronha-Blob L, Sacktor B, Farquhar MG (1984) Microdomains of distinctive composition in the kidney proximal tubule brush border. J Cell Biol 98: 1505–1513
Kerr MA, Kenny AJ (1974) The purification and specificity of a neutral endopeptidase from rabbit kidney brush border. Biochem J 137: 477–488
Kilpatrick DR, Srinivas RV, Stephens EB, Compans RW (1987) Effects of deletion of the cytoplasmic domain upon surface expression and membrane stability of a viral envelope glycoprotein. J Biol Chem 262: 16116–16121
Kleinman HK, Klebe RJ, Martin GR (1981) Role of collagenous matrices in the adhesion and growth of cells. J Cell Biol 88: 473–485
Klenk H-D, Choppin PW (1970a) Glycosphingolipids of plasma membranes of cultured cells and an enveloped virus (SV5) grown in these cells. Proc Natl Acad Sci USA 66: 57–64
Klenk H-D, Choppin PW (1970b) Plasma membrane lipids and paramyxovirus assembly. Virology 40: 939–947
Kollias G, Evans DJ, Ritter M, Beech J, Morris R, Grosveld F (1987) Ectopic expression of Thy-1 in the kidneys of transgenic mice induces functional and proliferative abnormalities. Cell 51: 21–31
Kondor-Koch C, Bravo R, Fuller SD, Cutler D, Garoff H (1985) Exocytic pathways exist to both apical and the basolateral surface of the polarized epithelial cell MDCK. Cell 43: 297–306
Lane EB (1982) Monoclonal antibodies provide specific intramolecular markers for the study of epithelial organization. J Cell Biol 92: 665–673
Laver WG, Valentine RC (1969) Morphology of isolated hemagglutinin and neuraminidase subunits of influenza virus. Virology 38: 105–119
Lazarowitz SG, Compans RW, Choppin PW (1971) Influenza virus structural and nonstructural proteins in infected cells and their plasma membranes. Virology 46: 830–843
LeBivic AL, Real FX and Rodriguez-Boulan E. (1989) Vectorial targeting of apical and basolateral plasma membrane proteins in a human adenocarcinoma cell line. Proc Natl Acad Sci USA 86: 9313–9317
Lisanti MP, Sargiacomo M, Graeve L, Saltiel AR, Rodriguez-Boulan E (1988) Polarized apical distribution of glycosyl-phosphatidylinositol-anchored proteins in a renal epithelial cell line. Proc Natl Acad Sci USA 85: 9557–9561
Lisanti MP, Bivic AL, Sargiacomo M, Rodriguez-Boulan E (1989) Steady state distribution and biogenesis of Madin-Darby canine kidney glycoproteins: evidence for intracellular sorting and polarized cell surface delivery. J Cell Biol 109:
Lodja Z (1974) Cytochemistry of enterocytes and of other cells in mucous membrane of the small intestine. In: Smith DA (ed) Biomembranes 4A Plenum, New York
Lombardi T, Montesano R, Wohlwend A, Amherdt M, Vassali J and Orci L (1985) Evidence for polarization of plasma membrane domains in pancreatic endocrine cells. Nature 313: 694–696
Louvard D, Godefroy O, Huet C, Sahuquillo-Merino C, Robine S, Coudrier E (1985) Basolateral membrane proteins are expressed at the surface of immature intestinal cell whereas transport of apical proteins is abortive. In: Gething M-J (ed) Current communications in molecular biology. Protein transport and secretion. Cold Spring Harbor Laboratory, New York, pp 168–173

Low MG (1987) Biochemistry of the glycosyl-phosphatidylinositol membrane protein anchors. Biochem J 244: 1–13

Ludens JH, Zimmerman WB, Schieders JR (1980) Nature of the inhibition of Cl^- transport by fucosemide: evidence for a direct effect on active transport in the toad cornea. Life Sci 27: 2453–2458

Madara JL, Dharmsathaphorn K (1985) Occluding junction structure-function relationship in a cultured epithelial monolayer. J Cell Biol 101: 2124–2133

Maratje GV, Nash B, Haschemeyer RH, Tate SS (1979) Ultrastructural localization of γ-glutamyl transferase in rat kidney and jejunum. FEBS Lett 107: 436–440

Martinez-Palomo AI, Erlij D (1975) Structure of tight junctions in epithelia with different permeability. Proc Natl Acad Sci USA 72: 4487–4491

Martinez-Palomo AI, Meza I, Beaty G, Cereijido M (1980) Experimental modulation of occluding junctions in a cultured transporting epithelium. J Cell Biol 87: 736–745

Marxer A, Stieger B, Quaroni A, Kashgarian M Hauri H (1989) (Na^+ + K^+)-ATPase and plasma membrane polarity of the intestinal epithelial cells: presence of a brush border antigen in the distal large intestine that is immunologically related to the β-subunit. J Cell Biol 109: 1057–1069

Matlin KS (1986a) Ammonium chloride slows transport of influenza hemagglutinin but does not lead to missorting in MDCK cells. J Biol Chem 261: 15172–15178

Matlin KS (1986b) The sorting of proteins to the plasma membrane in epithelial cells. J Cell Biol 103: 2565–2568

Matlin K, Simons K (1984) Sorting of an apical plasma membrane glycoprotein occurs before it reaches the cell surface in cultured epithelial cells. J Cell Biol 99: 2131–2139

Matter K, Brauchbar M, Bucher K, Hauri H-P (1990). Sorting of endogenous plasma membrane proteins occurs from two sites in cultured human intestinal epithelial cells (CaCo-2). Cell 60: 429–437

Mayne J, Compans RW (1988) Somatic cell mutants of Madin-Darby canine kidney cells: cells defective in the proper maturation of VSV G protein. Abst J Cell Biol 107: 131a

McNutt NS, Weinstein RS (1973) Membrane ultrastructure at mammalian intercellular junctions Progr Biophys Mol Biol 26: 45–102

McQueen NL, Nayak DP, Jones LV, Compans RW (1984) Chimeric influnza virus hemagglutinin containing either the NH2 terminus or the COOH terminus of G protein of vesicular stomatitis virus is defective in transport to the cell surface. Proc Natl Acad Sci USA 81: 395–399

McQueen NL, Nayak DP, Stephens EB, Compans RW (1986) Polarized expression of a chimeric protein in which the transmembrane and cytoplasmic domains of the influenza virus hemagglutinin have been replaced by those of vesicular stomatitis virus G protein. Proc Natl Acad Sci USA 83: 9318–9322

McQueen NL, Nayak DP. Stephens EB, Compans RW (1987) Basolateral expression of a chimeric protein in which the transmembrane and cytoplasmic domains of vesicular stomatitis virus G protein have been replaced by those of the influenza virus hemagglutinin. J Biol Chem 262: 16233–16240

Mege R-M, Matsuzaki F, Gallin W, Goldberg JI, Cunningham BA, Edelman GM (1988) Construction of epithelioid sheets by transfection of mouse sarcoma cells with cDNAs for chicken cell adhesion molecules. Proc Natl Acad Sci USA 85: 7274–7278

Meier PJ, Sztul Es, Reuben A, Boyer JL (1984) Structural and functional polarity of canalicular and basolateral plasma membrane vesicles isolated in high yield from rat liver. J Cell Biol 98: 991–1000

Melancon P, Glick BS, Malhotra V, Weidman PJ, Sarafini T, Gleason M, Orci L, Rothman J (1987) Involvement of GTP binding "G" proteins in transport through the Golgi stack Cell. 51: 1053–1062

Mestecky J, McGhee JR (1987) Immunoglobulin A (IgA): Molecular and cellular interactions involved in IgA biosynthesis and immune response. Adv Immunol 40: 153–245

Meza I, Ibarra G, Saanero M, Martinez-Palomo A, Cereijido M (1980) Occluding junctions and cytoskeletal components in a transporting epithelium. J Cell Biol 87: 746–754

Milhorat TH, Davis DA, Hammock MK (1975) Localization of ouabain-sensitive Na^+-K^+-ATPase in frog, rabbit and rat choroid plexus. Brain Res 99: 170–174

Mircheff AK, Sachs G, Hanna SD, Labiner CS, Rabon E, Douglas AP, Walling MW, Wright EM (1979) Highly purified basal lateral plasma membranes from rat duodenum. Physical criteria for purity. J Membr Biol 50: 343–363

Misek DE, Bard E, Rodriguez-Boulan E (1984) Biogenesis of epithelial cell polarity: Intracellular sorting and vectorial exocytosis of an apical plasma membrane glycoprotein. Cell 39: 537–546

Molitoris BA, Simon FR (1985) Renal cortical brush border and basolateral membranes: cholesterol and phospholipid composition and relative turnover. J Membr Biol 83: 207–215

Molitoris BA, Simon FR (1986) Maintenance of epitelial surface membrane lipid polarity: a role for differing phospholipid translocation rates. J Membr Biol 94: 47–53
Mostov KE, Blobel G (1983) Transcellular transport of polymeric immunoglobulins by secretory components: a model system for studying intracellular protein sorting Ann NY Acad Sci 409: 441–451
Mostov KE, Deitcher DL. (See original text)
Mostov KE, Simister NE (1985) Transcytosis. Cell 43: 389–390
Mostov KE, Breitfeld P, Harris JM (1987) An anchor-minus form of the polymeric immunoglobulin receptor is secreted predominantly apically in Madin-Darby canine kidney cells. J Cell Biol 105: 2031–2036
Murer H, Kinne R (1980) The use of isolated membrane vesicles to study transport processes. J Membr Biol 55: 81–95
Novick P, Field C, Scheckman R (1980) Identification of 23 complementation groups required for post-translational events in yeast secretory pathway. Cell 21: 205–215
Owens RJ, Compans RW (1989) Expression of the human immunodeficiency virus envelope glycoprotein is restricted to basolateral surfaces of polarized epithelial cells. J Virol 63: 978–982
Pascher I (1976) Molecular arrangements in sphingolipids. Conformation and hydrogen bonding of ceramide and their implication on membrane stability and permeability Biochim Biophys Acta 455: 433–451
Pavelka M, Ellinger A, Gangl A (1983) Effect of colchicine on rat small intestinal absorptive cells. I Formation of basolateral microvillus borders. J Ultrastruct Res 85: 249–259
Pfeffer SR, Rothman JE (1987) Biosynthetic protein transport and sorting by the endoplasmic reticulum and Golgi. Ann Rev Biochem 56: 829–852
Pfeiffer S, Fuller SD, Simons K (1985) Intracellular sorting and basolateral appearance of the G protein of vesicular stomatitis virus in MDCK cells. J Cell Biol 101: 470–476
Pinto da Silva P, Kachar B (1982) On tight junction structure. Cell 28: 441–450
Pisan M, Ripoche P (1976) Redistribution of surface molecules in dissociated epithelial cells. J Cell Biol 71: 907–20
Puddington L, Woodgett, Rose JK (1987) Replacement of the cytoplasmic domain alters sorting of a viral glycoprotein in polarized cells. Proc Natl Acad Sci USA 84: 2756–2760
Quaroni A, Isselbacher KJ (1981) Study of intestinal cell differentiation with monoclonal antibodies to intestinal cell surface components. Dev Biol 111: 267–279
Quaroni A, Kirsch K, Weiser MM (1979a) Synthesis of membrane glycoproteins in rat small-intestinal villus cells. Effect of colchicine on the redistribution of *L*-(1,5,6,-3H)-fucose-labelled membrane glycoproteins among Golgi, lateral, basal, and microvillus membranes. Biochem J 182: 213–221
Quaroni A, Kirsh K, Weiser MM (1979b) Synthesis of membrane glycoproteins in rat small-intestinal villus cells. Redistribution of *L*-(1,5,6,-3H)-fucose-labelled membrane glycoproteins among Golgi, lateral, basal, and microvillus membranes *in vivo*. Biochem J 182: 203–212
Rabito CA, Kreisberg JI, Wight D (1984) Alkaline phosphatase and γ-glutamyl transpeptidase as polarization markers during the organization of LLC-PK cells into an epithelial membrane. J Biol Chem 259: 574–582
Reik L, Petzgold GL, Higgins JA, Greengard P, Bannett RJ (1970) Hormone sensitive adenyl cyclase: cytochemical localization in liver. Scince 168: 382–386
Reuss L, Finn A (1974) Passive electrical properties of toad urinary bladder epithelium J Gen Physiol 64: 1–25
Revel JP, Karnovsky MJ (1967) Hexagonal array of subunits in intercellular junctions of the mouse heart and liver. J Cell Biol 33: C7–C12
Richardson JCW, Scalera V, Simmons NL (1981) Identification of two strains of MDCK cells which resemble separate nephron tubule segments. Biochim Biophys Acta 673: 26-36
Rindler MJ, Ivanov IE, Plesken H, Rodriguez-Boulan E, Sabatini DD (1984) Viral glycoproteins destined for apical or basolateral plasma membrane domains traverse the same Golgi apparatus during their transport in doubly infected MDCK cells. J Cell Biol 98: 1304–1319
Rindler MJ, Ivanov IE, Plesken H, Sabatini DD (1985) Polarized delivery of viral glycoproteins to the apical and basolateral plasma membranes of MDCK cells infected with temperature sensitive viruses. J Cell Biol 104: 231-241
Rindler MJ, Ivanov IE, Sabatini DD (1987) Microtubule-acting drugs lead to the nonpolarized delivery of influenza hemagglutinin to the cell surface of polarized MDCK cells. J Cell Biol 104: 231–241
Rindler MJ, Traber MG (1988) Polarized secretion of newly synthesized proteins by cultured intestinal epithelial cells: a basolaterally directed default pathway. J Cell Biol 107: 471-479
Rodriguez-Boulan E (1983) Membrane biogenesis, enveloped RNA viruses, and epithelial polarity. In: Satir BH (ed) Modern cell biology, vol 1. Liss, New York, pp 119–170

Rodriguez-Boulan E, Nelson WJ (1989) Morphogenesis of the polarized epithelial cell phenotype. Science 245: 718–724

Rodriguez-Boulan E, Pendergast M (1980) Polarized distribution of viral envelope proteins in the plasma membrane of infected epithelial cells. Cell 20: 45–54

Rodriguez-Boulan E, Sabatini DD (1978) Asymmetric budding of viruses in epithelial cell monolayers: a model system for study of epithelial polarity. Proc Natl Acad Sci USA 75: 7051–5075

Rodriguez-Boulan E, Paskiet KT, Sabatini DD (1983) Assembly of enveloped viruses in MDCK cells: polarized budding from single attached cells and from clusters of cells in suspension. J Cell Biol 96: 866–874

Rodriguez-Boulan E, Paskiet KT, Salas PJ, Bard E (1984) Intracellular transport of influenza virus hemagglutinin to the apical surface of MDCK cells. J Cell Biol 98: 308–319

Roise D, Horvath SJ, Tomich JM, Richards JH, Schatz G. (1986) A chemically synthesized presequence of an imported mitochondrial protein can form an amphipathic helix and perturb natural and artificial phospholipid bilayers. EMBO J 5: 1327–1334

Roman LM, Garoff H (1986) Alteration of the cytoplasmic domain of the membrane spanning glycoprotein p62 of Semliki Forest virus does not affect its polar distribution in established cell lines of Madin-Darby canine kidney cells. J Cell Biol 103: 2607–2618

Roman LM, Hubbard AL (1983) A domain specific marker for the hepatocyte plasma membrane; localization of leucine aminopeptidase to the bile canalicular domain. J Cell Biol 96: 1548–1558

Roman LM, Hubbard AL (1984a) A domain specific marker for the hepatocyte plasma membrane. II. Ultrastructural localization of leucine aminopeptidase to the bile canalicular domain of isolated rat liver plasma membranes. J Cell Biol 98: 1488–1497

Roman LM, Hubbard AL (1984b) A domain specific marker for the hepatocyte plasma membrane. III. Isolation of bile canalicular membrane by immunoadsorption. J Cell Biol 98: 1498–1504

Rose JK, Gallione CJ. (1981) Nucleotide sequences of the mRNAs encoding the vesicular stomatitis G and M proteins as determined from cDNA clones containing complete coding regions. J Virol 39: 519–528

Roth MG, Compans RW (1981) Delayed appearance of pseudotypes between vesicular stomatitis virus and influenza during mixed infection of MDCK cells. J Virol 40: 848–860

Roth MG, Fitzpatric JP, Compans RW (1979) Polarity of influenza and vesicular stomatitis virus in maturation in MDCK cells: lack of a requirement for glycosylation of viral glycoproteins. Proc Natl Acad Sci USA 76: 6430–6434

Roth MG, Compans RW, Giusti L,Davis AR, Nayak DP, Gething M-J, Sambrook J (1983a) Influenza hemagglutinin expression is polarized in cells infected with recombinant SV40 viruses carrying cloned hemagglutinin DNA. Cell 33: 435–443

Roth MG, Srinivas RV, Compans RW (1983b) Basolateral maturation of retroviruses in polarized epithelial cells. J Virol 45: 1065–1073

Roth MG, Gundersen D, Patil N, Rodriguez-Boulan E (1987) The large external domain is sufficient for the correct sorting of secreted or chimeric influenza virus hemagglutinins in polarized monkey kidney cells. J Cell Biol 104: 769–782

Rothman JE, Tsai DK, Dawidowicz EA, Lenard J (1976) Transbilayer phospholipid asymmetry and its maintenance in the membrane of influenza virus. Biochemistry 15: 2361–2370

Sabitini DD, Griepp EB, Rodriguez-Boulan E, Dolan WJ, Robbins ES, Papadopoulous S, Ivanov IE, Rindler MJ (1983) Biogenesis of epithelial surface polarity. In: Satir BH (ed) Modern cell biology, vol 2. Liss, New York, pp 419–450

Salas PJI, Misek DE, Vega-Sala DE, Gunderson D, Cereijido M, Rodriguez-Boulan E (1986) Microtubules and actin filaments are not critically involved in the biogenesis of the epithelial cell surface polarity. J Cell Biol 102: 1853–1867

Salminen A, Novick P. (1987) A *ras*-like protein is required for a post-Golgi event in yeast secretion. Cell 49: 527–538

Santos E, Nebrada ER (1989) Structural and functional properties of ras proteins. FASEB J 3: 2151–2162

Schmitt HD, Puzicha M, Gallwitz D (1988) Study of a temperature sensitive mutant of the *ras*-related *YPT1* gene product in yeast suggests a role in the regulation of intracellular calcium. Cell 53: 635–647

Schwartz IL, Shaltz LJ, Kinne-Saffran E, Kinne R (1974) Target cell polarity and membrane phosphorylation in relation to the mechanism of action of antidiuretic hormone. Proc Natl Acad Sci USA 71: 2595–2599

Segev N, Mulholland J, Botstein D (1988) The yeast GTP-binding protein YPT1 and a mammalian counterpart are associated with the secretion machinery. Cell 52: 915–924

Semenza G (1976) Small intestinal disaccharidases: their properties and role as sugar translocators across natural and artificial membranes. In: Membrane transport, enzymes of biological membranes, Vol 3, pp 349–382. New York: Plenum
Simons K, Fuller SD (1985) Cell surface polarity in epithelia, Ann Rev Cell Biol 1: 243–288
Simons K, Garoff H (1980) The budding mechanism of enveloped animal viruses. J Gen Virol 50: 1–21
Simons K, Virta H (1987) Perforated MDCK cells support intracellular transport. EMBO J 6: 2241–2247
Skehel JJ (1972) Polypeptide synthesis in influenza virus-infected cells. Virology 49: 23–36
Spiegel S, Blumenthal R, Fishman PH, Handler JS (1985) Gangliosides do not move from apical to basolateral plasma membrane in cultured epithelial cells. Biochim Biophys Acta 821: 310–318
Srinivas RV, Balachandran N, Alonso-Caplen FV, Compans RW (1986) Expression of herpes simplex virus glycoproteins in polarized epithelial cells. J Virol 58: 689–693
Stephens EB, Compans RW (1986) Nonpolarized expression of a secreted murine leukemia virus glycoprotein in polarized epithelial cells. Cell 47: 1053–1059
Stephens EB, Compans RW (1988) Assembly of animal viruses at cellular membranes. Ann Rev Microbiol 42: 489–516
Stephens EB, Compans RW, Earl P, Mos B (1986) Surface expression of viral glycoproteins is polarized in epithelial cells infected with recombinant vaccinia viral vectors. EMBO J 5: 237–245
Stern CD, MacKenzie DO (1983) Sodium transport and control of epithelial polarity in the early chick embryo. J Embryol Exp Morphol 77: 73–98
Stevenson BR, Siliciano JD, Moosekar MS, Goodenough DA (1986) Identification of ZO-1: a high molecular weight polypeptide associated with the tight junction (zona occludens) in a variety of epithelia. J Cell Biol 103: 755–766
Strous GJAM, Willemsen R, van Kerkhof P, Slot JW, Geuze HJ, Lodish HF (1983) Vesicular stomatitis virus glycoprotein, albumin, and transferrin are transported to the cell surface via the same Golgi vesicles. J Cell Biol 97: 1815–1822
Sugrue SP, Hay ED (1981) Response of basal epithelial cell surface and cytoskeleton to solubilized extracellular matrix molecules. J Cell Biol 91: 45–54
Takemura S, Omiro K, Tanaka K, Omori K, Matsuura S, Tashito Y (1984) Quantitative immunoferritin localization of Na^+/K^+ ATPase on canine hepatocyte cell surface. J Cell Biol 99: 1502–1510
Tartakoff AM, Vasali P (1977) Plasma cell immunoglobulin secretion: arrest is accompanied by alterations in the Golgi complex. J Exp Med 146: 1332–1345
Taub M (1985) Tissue culture of epithelial cells. Plenum, New York
Thompson TE, Tillack TW (1985) Organization of glycosphingolipids in bilayers and plasma membranes of mammalian cells. Ann Rev Biophys Chem 14: 361–386
Thompson TE, Barenholz Y, Brown RE, Correa-Freire M, Young WW, Tillack TW (1986) In: Freysz L (ed) Enzymes of lipid metabolism. Plenum, New York, pp 387–396
Tilney L (1983) Interactions between actin filaments and membranes give spatial organization to cells. In: Satir BH (ed) Modern cell biology, Vol 2. Liss, New York, pp 63–199
Timple R, Martin GR (1982) Components of basal membranes. In: Furthmayr H (ed) Immunocytochemistyr of the extracellular matrix. CRC Press, Boca Raton
van Meer G, Simons K (1982) Viruses budding from either the apical or basolateral domain of MDCK cells have unique phospholipid composition. EMBO J 1: 847–852
van Meer G, Simons K (1986) The functions of tight junctions in maintaining differences in lipid composition between the apical and basolateral cell surface domains of MDCK cells. EMBO J 5: 1455–1464
van Meer G, Simons K (1988) Lipid polarity and sorting in epithelial cells. J Cell Biochem 36: 51–58
van Meer G, Stelzer EHK, Wijnaendts-van-Resandt RW, Simons K (1987) Sorting of sphingolipids in epithelial (MDCK) cells. J Cell Biol 105: 1623–1635
Varghese JN, Laver WG, Colman PM (1983) Structure of the influenza virus glycoprotein neuraminidase at 2.9 Å resolution Nature 303: 35–40
Vega-Salas DE, Salas PJI, Rodriguez-Boulan E (1987) Modulation of the expression of an apical plasma membrane protein of MDCK epithelial cells: cell-cell interactions control the appearance of novel intracellular storage compartment. J Cell Biol 104: 1249–1259
Vega-Salas DE, Salas PJI, Rodriguez-Boulan E (1988) Exocytosis of vacuolar apical compartment (VAC): a cell-cell contact controlled mechanism for the establishment of the apical plasma membrane domain in epithelial cells. J Cell Biol 107: 1717–1728
Volk T, Geiger B (1984) A 135-kd membrane protein of intercellular adherens junctions. EMBO J 3: 2249–2260
Von Heijne G (1985) Signal sequences: The limits of variation. J Mol Biol 184: 99–105

von der Mark R, Mollenhauer J, Kuhl K, Bee J, Lesot H (1984) Anchorins: a new class of membrane proteins involved in cell-matrix interactions. In. Trelstadt RL (ed) The role of extracellular matrix in development. Liss, New York, pp 67–87

Wagner RR (1987) The rhabdoviruses. Plenum, New York

Warren SL, Handell LM, Nelson WL (1988) Elevated expression of pp60$^{c\text{-}src}$ alters a selective morphogenetic property of epithelial cells in vitro without a mitogenic effect. Mol Cell Biol: 632–646

Weiss R, Teich N, Varmus H, Coffin J (1984) RNA tumor viruses, Molecular biology of tumor viruses 2nd ed. 2. Supplements/appendixes. Cold Spring Harbor Laboratory, New York

Wieland FT, Gleason ML, Serafini TA, Rothman JE (1987) The rate of bulk flow from the endoplasmic reticulum to the cell surface. Cell 50: 289–300

Wiley DC, Skehel JJ (1987) The structure and function of the hemagglutinin membrane glycoprotein of influenza virus. Ann Rev Biochem 56: 365–394

Wiley DC, Skehel JJ, Waterfield MD (1977) Evidence from studies with a cross-linking agent that the hemagglutinin of influenza virus is a trimer.Virology 79: 446–448

Wilson IA, Skehel JJ, Wiley DV (1981) Structure of the hemagglutinin membrane glycoprotein of influenza virus at 3Å resolution. Nature 289: 366–373

Wisher MH, Evans WH (1975) Functional polarity of rat hepatocyte surface membrane: isolation and characterization of plasma membrane subfractions from the blood-sinusoidal, bile-canalicular and contiguous surfaces of the hepatocyte. Biochem J 146: 375–388

Woodman PDG, Edwardson JM (1986) A cell free assay for the insertion of a viral glycoprotein into the plasma membrane. J Cell Biol 103: 1829–1835

Zahraoui A, Touchot N, Chardin P, Tavitian A (1989) The human *rab* gene encode a family of GTP binding proteins related to the yeast YPT1 and Sec4 products involved in secretion. J Biol Chem 264: 12394–12401

Ziomeck CA, Schulman S, Eddin M (1980) Redistribution of membrane proteins in isolated mouse epithelial cells. J Cell Biol 86: 849–857

Subject Index

Current Topics in Microbiology and Immunology

Volumes published since 1986 (and still available)

Vol. 125: **Wu, Henry C.; Tai, Phang C. (Ed.):** Protein Secretion and Export in Bacteria. 1986. 34 figs. X, 211 pp. ISBN 3-540-16593-2

Vol. 126: **Fleischer, Bernhard; Reimann, Jörg; Wagner, Hermann (Ed.):** Specificity and Function of Clonally Developing T-Cells. 1986. 60 figs. XV, 316 pp. ISBN 3-540-16501-0

Vol. 127: **Potter, Michael; Nadeau, Joseph H.; Cancro, Michael P. (Ed.):** The Wild Mouse in Immunology. 1986. 119 figs. XVI, 395 pp. ISBN 3-540-16657-2

Vol. 128: 1986. 12 figs. VII, 122 pp. ISBN 3-540-16621-1

Vol. 129: 1986. 43 figs., VII, 215 pp. ISBN 3-540-16834-6

Vol. 130: **Koprowski, Hilary; Melchers, Fritz (Ed.):** Peptides as Immunogens. 1986. 21 figs. X, 86 pp. ISBN 3-540-16892-3

Vol. 131: **Doerfler, Walter; Böhm, Petra (Ed.):** The Molecular Biology of Baculoviruses. 1986. 44 figs. VIII, 169 pp. ISBN 3-540-17073-1

Vol. 132: **Melchers, Fritz; Potter, Michael (Ed.):** Mechanisms in B-Cell Neoplasia. Workshop at the National Cancer, Institute, National Institutes of Health, Bethesda, MD, USA, March 24–26, 1986. 1986. 156 figs. XII, 374 pp. ISBN 3-540-17048-0

Vol. 133: **Oldstone, Michael B. (Ed.):** Arenaviruses. Genes, Proteins, and Expression. 1987. 39 figs. VII, 116 pp. ISBN 3-540-17246-7

Vol. 134: **Oldstone, Michael B. (Ed.):** Arenaviruses. Biology and Imminotherapy. 1987. 33 figs. VII, 242 pp. ISBN 3-540-14322-6

Vol. 135: **Paige, Christopher J.; Gisler, Roland H. (Ed.):** Differentiation of B Lymphocytes. 1987. 25 figs. IX, 150 pp. ISBN 3-540-17470-2

Vol. 136: **Hobom, Gerd; Rott, Rudolf (Ed.):** The Molecular Biology of Bacterial Virus Systems. 1988. 20 figs. VII, 90 pp. ISBN 3-540-18513-5

Vol. 137: **Mock, Beverly; Potter, Michael (Ed.):** Genetics of Immunological Diseases. 1988. 88 figs. XI, 335 pp. ISBN 3-540-19253-0

Vol. 138: **Goebel, Werner (Ed.):** Intracellular Bacteria. 1988. 18 figs. IX, 179 pp. ISBN 3-540-50001-4

Vol. 139: **Clarke, Adrienne E.; Wilson, Ian A. (Ed.):** Carbohydrate-Protein Interaction. 1988. 35 figs. IX, 152 pp. ISBN 3-540-19378-2

Vol. 140: **Podack, Eckhard R. (Ed.):** Cytotoxic Effector Mechanisms. 1989. 24 figs. VIII, 126 pp. ISBN 3-540-50057-X

Vol. 141: **Potter, Michael; Melchers, Fritz (Ed.):** Mechanisms in B-Cell Neoplasia 1988. Workshop at the National Cancer Institute, National Institutes of Health, Bethesda, MD, USA, March 23–25, 1988. 1988. 122 figs. XIV, 340 pp. ISBN 3-540-50212-2

Vol. 142: **Schüpach, Jörg:** Human Retrovirology. Facts and Concepts. 1989. 24 figs. 115 pp. ISBN 3-540-50455-9

Vol. 143: **Haase, Ashley T.; Oldstone Michael B. A. (Ed.):** In Situ Hybridization 1989. 22 figs. XII, 90 pp. ISBN 3-540-50761-2

Vol. 144: **Knippers, Rolf; Levine, A. J. (Ed.):** Transforming. Proteins of DNA Tumor Viruses. 1989. 85 figs. XIV, 300 pp. ISBN 3-540-50909-7

Vol. 145: **Oldstone, Michael B. A. (Ed.):** Molecular Mimicry. Cross-Reactivity between Microbes and Host Proteins as a Cause of Autoimmunity. 1989. 28 figs. VII, 141 pp. ISBN 3-540-50929-1

Vol. 146: **Mestecky, Jiri; McGhee, Jerry (Ed.):** New Strategies for Oral Immunization. International Symposium at the University of Alabama at Birmingham and Molecular Engineering Associates, Inc. Birmingham, AL, USA, March 21–22, 1988. 1989. 22 figs. IX, 237 pp. ISBN 3-540-50841-4

Vol. 147: **Vogt, Peter K. (Ed.):** Oncogenes. Selected Reviews. 1989. 8 figs. VII, 172 pp. ISBN 3-540-51050-8

Vol. 148: **Vogt, Peter K. (Ed.):** Oncogenes and Retroviruses. Selected Reviews. 1989. XII, 134 pp. ISBN 3-540-51051-6

Vol. 149: **Shen-Ong, Grace L. C.; Potter, Michael; Copeland, Neal G. (Ed.):** Mechanisms in Myeloid Tumorigenesis. Workshop at the National Cancer Institute, National Institutes of Health, Bethesda, MD, USA, March 22, 1988. 1989. 42 figs. X, 172 pp. ISBN 3-540-50968-2

Vol. 150: **Jann, Klaus; Jann, Barbara (Ed.):** Bacterial Capsules. 1989. 33 figs. XII, 176 pp. ISBN 3-540-51049-4

Vol. 151: **Jann, Klaus; Jann, Barbara (Ed.):** Bacterial Adhesins. 1990. 23 figs. XII, 192 pp. ISBN 3-540-51052-4

Vol. 152: **Bosma, Melvin J.; Phillips, Robert A.; Schuler, Walter (Ed.):** The Scid Mouse. Characterization and Potential Uses. EMBO Workshop held at the Basel Institute for Immunology, Basel, Switzerland, February 20–22, 1989. 1989. 72 figs. XII, 263 pp. ISBN 3-540-51512-7

Vol. 153: **Lambris, John D. (Ed.):** The Third Component of Complement. Chemistry and Biology. 1989. 38 figs. X, 251 pp. ISBN 3-540-51513-5

Vol. 154: **McDougall, James K. (Ed.):** Cytomegaloviruses. 1990. 58 figs. IX, 286 pp. ISBN 3-540-51514-3

Vol. 155: **Kaufmann, Stefan H. E. (Ed.):** T-Cell Paradigms in Parasitic and Bacterial Infections. 1990. 24 figs. IX, 162 pp. ISBN 3-540-51515-1

Vol. 156: **Dyrberg, Thomas (Ed.):** The Role of Viruses and the Immune System in Diabetes Mellitus. 1990. 15 figs. XI, 142 pp. ISBN 3-540-51918-1

Vol. 157: **Swanstrom, Ronald; Vogt, Peter K. (Ed.):** Retroviruses. Strategies of Replication. 1990. 40 figs. XII, 260 pp. ISBN 3-540-51895-9

Vol. 158: **Muzyczka, Nicholas (Ed.):** Viral Expression Vectors. 1991. Approx. 20 figs. Approx. XII, 190 pp. ISBN 3-540-52431-2

Vol. 159: **Gray, David; Sprent, Jonathan (Ed.):** Immunological Memory. 1990. 38 figs. XII, 156 pp. ISBN 3-540-51921-1

Vol. 160: **Oldstone, Michael B. A.; Koprowski, Hilary (Ed.):** Retrovirus Infections of the Nervous System. 1990. 16 figs. XII, 176 pp. ISBN 3-540-51939-4

Vol. 161: **Racaniello, Vincent R. (Ed.):** Picornaviruses. 1990. 12 figs. X, 194 pp. ISBN 3-540-52429-0

Vol. 162: **Roy, Polly; Gorman, Barry M. (Ed.):** Bluetongue Viruses. 1990. 37 figs. X, 200 pp. ISBN 3-540-51922-X

Vol. 163: **Turner, Peter C.; Moyer, Richard W. (Ed.):** Poxviruses. 1990. 23 figs. X, 210 pp. ISBN 3-540-52430-4

Vol. 164: **Bækkeskov, Steinnun; Hansen, Bruno (Ed.):** Human Diabetes. 1990. 9 figs. X, 198 pp. ISBN 3-540-52652-8

Vol. 165: **Bothwell, Mark (Ed.):** Neuronal Growth Factors. 1991. 14 figs. IX, 173 pp. ISBN 3-540-52654-4

Vol. 166: **Potter, Michael; Melchers, Fritz (Ed.):** Mechanisms in B-Cell Neoplasia 1990. 143 figs. XIX, 380 pp. ISBN 3-540-52886-5

Vol. 167: **Kaufmann, Stefan H. E. (Ed.):** Heat Shock Proteins and Immune Response. 1991. 18 figs. IX, 214 pp. ISBN 3-540-52857-1

Vol. 168: **Mason, William S.; Seeger, Christoph (Ed.):** Hepadnaviruses. Molecular Biology and Pathogenesis. 1991. 21 figs. X, 206 pp. ISBN 3-540-53060-6

Vol. 169: **Kolakofsky, Daniel (Ed.):** Bunyaviridae. 1991. 34 figs. X, 256 pp. ISBN 3-540-53061-4

MIX
Papier aus verantwortungsvollen Quellen
Paper from responsible sources
FSC® C105338
FSC
www.fsc.org

If you have any concerns about our products,
you can contact us on
ProductSafety@springernature.com

In case Publisher is established outside the EU,
the EU authorized representative is:
Springer Nature Customer Service Center GmbH
Europaplatz 3, 69115 Heidelberg, Germany

Printed by Libri Plureos GmbH
in Hamburg, Germany